ROBERT P. CREASE is a professor in the department of
Philosophy at Stony Brook University in New York, and historian
at Brookhaven National Laboratory. He writes a monthly column,
"Critical Point," for *Physics World* magazine. His previous books
include *Making Physics: A Biography of Brookhaven National
Laboratory; The Play of Nature: Experimentation as Performance;
The Second Creation: Makers of the Revolution in Twentieth-Century
Physics* (with Charles C. Mann); and—with Robert Serber—
Peace & War: Reminiscences of a Life on the Frontiers of Science.
Crease's translations include *American Philosophy of Technology:
The Empirical Turn.* He lectures widely, and his articles and reviews
have appeared in *The Atlantic Monthly, The New York Times
Magazine, The Wall Street Journal, Smithsonian,*
and elsewhere. He lives in New York City.

The Prism and the Pendulum

RANDOM HOUSE TRADE PAPERBACKS

New York

The Prism and the Pendulum

THE TEN MOST BEAUTIFUL
EXPERIMENTS IN SCIENCE

Robert P. Crease

Library of Congress Cataloging-in-Publication Data

Crease, Robert P.
The prism and the pendulum: the ten most beautiful experiments in
science / Robert P. Crease.
p. cm.
Includes bibliographical references.
ISBN 0-8129-7062-4
1. Science—History. 2. Science—Experiments. I. Title.
Q125.C67 2003   509—dc21   2003054765

Book design by Barbara M. Bachman

To wild things everywhere

Contents

—

List of Illustrations

——

The Moment of Transition

—

I CAN'T REMEMBER THE FIRST TIME I HEARD scientists refer to an experiment as "beautiful," but I do remember the first time I grasped what they were talking about.

Many years ago I was sitting in a dimly lit academic office in the physics building at Harvard University, surrounded by untidy heaps of books and papers. Across from me was Sheldon Glashow, an energetic physicist whose features, including his industrial-thickness glasses, were almost hidden behind an oracular veil of cigar smoke. "That was a beautiful experiment," he was saying. "An absolutely *beautiful* experiment!" Something about his intensity and emphasis suddenly made me realize that he was choosing his words carefully. In his eyes, the experiment he was describing was—quite literally—a thing of beauty.

Glashow is not an uncultivated person. Like many scientists, he knows much more about the arts and humanities than the practitioners of those fields generally know about his own: high-energy physics. He is, moreover, an outstanding scientist, and had been awarded the Nobel Prize in Physics in 1979, a few years before our conversation. At that moment in his office I was forced to consider the possibility that someone could truly regard a scientific experiment as beautiful, and mean much the same thing that most of us do when we call a landscape, person, or painting beautiful.

I was curious to learn more about the particular experiment that had excited Glashow, which he had referred to in scientific short-hand as the "SLAC neutral currents" experiment. It was a difficult and complex undertaking that had consumed the efforts of scientists, engineers, and technicians over several years. It had taken nearly ten years to plan and build, and was finally carried out in the spring of 1978 at a two-mile-long particle accelerator at the Stanford Linear Accelerator Center (SLAC), south of San Francisco in the Santa Clara mountains. The experiment involved creating polarized electrons—electrons with spins oriented in the same direction—and then shooting them down the accelerator at nearly the speed of light, slamming them into a batch of protons and neutrons, and observing the results. At stake was a comprehensive new theory about the structure of matter at its most fundamental levels—a theory that Glashow had been instrumental in developing. If the theory were correct, the experimenters should see a slight difference between the way the electrons ricocheted off the proton targets when the electrons were polarized in opposite directions, indicating the presence of something called "parity-violating neutral currents." An extremely slight difference—about one per ten thousand electrons. Observing it would require so much precision—and for the experiment to be convincing, the scientists would have to watch *ten billion* electrons—that many scientists thought it impossible, or that the outcome would be inconclusive.

But a few days into the experiment, it became clear that the answer was not in the least ambiguous or questionable, and that the ambitious theory was correct. (Glashow and two other scientists would win Nobel Prizes for their roles in the creation of this theory.) Indeed, the superbly executed experiment made the existence of a new fundamental feature of nature—parity-violating neutral currents—so vividly apparent to anyone trained in physics that even those who did not participate found the work moving. When one of the scientists involved described the experimental work and its results to outsiders for the first time, in a talk held at the accelerator's auditorium in June 1978, it was the first instance that anyone in the audience could remember when nobody in the laboratory audience—usually a contentious bunch—challenged the results. Indeed, there were no questions at all. Those present also recall that the applause following the talk was longer, more appreciative, more respectful than usual.[1]

The idea of beautiful experiments made me wonder what else belonged on a list of beautiful experiments. And it raised issues that intrigued me from both poles of my dual career as a philosopher and as a historian of science: What does it mean for experiments, if they can be beautiful? And what does it mean for beauty, if experiments can possess it?

WHEN I TALK TO nonscientists about the beauty of experiments, they are often skeptical. There are three factors at work, I think, in this skepticism. One is social: When scientists present themselves in public—reporting their work formally or talking to journalists—they rarely use the word "beauty." The social convention for researchers is to appear as objective observers of nature, and to downplay the subjective and personal. To fulfill this image, scientists portray experiments as purely functional, as merely the manipulation of a set of instruments that almost automatically produce correct data.

A second factor is cultural: the way science is often taught in schools. Textbooks use experiments as vehicles for the lesson plan, as a prop for getting students to acquire deeper understanding. Viewing experiments as hurdles to pass a course, students can easily overlook their beauty.

A third factor is the philosophical prejudice that true beauty can only be found in the abstract. "Euclid alone has looked on Beauty bare," the poet Edna St. Vincent Millay declared. For this reason, discussions of beauty in science usually focus on its role in theories and explanations. Abstractions, like equations, models, and theories, are said to possess simplicity, clarity, insight, depth, timelessness, and other properties that we tend to associate with beauty. Experiments—messing about with machines, hardware, chemicals, and organisms—do not seem to fit the bill.

Working scientists know that on the laboratory floor experiments are mainly tedious work. Most of a scientist's time is spent calibrating, preparing, designing, ironing out glitches, solving routine problems, begging for money and support. Much of science consists of extending only incrementally what we can do, or what we know. But now and then, unpredictably yet inevitably, an event occurs that crystallizes a new insight and reshapes how we perceive things. It pulls us out of a confused state to show us—directly, and without further question—what is important, transforming our ideas about nature. Scientists tend to call such moments "beautiful."

The word crops up in talks, memos, letters, interviews, notebooks, and the like. "Beauty. Publish this surely, beautiful!" the Nobel Prize–winning physicist Robert Millikan wrote on a page of his lab notebook in 1912—though he did not use the word "beauty" in the scientific paper he subsequently published. James Watson, seeing Rosalind Franklin's now famous photograph of the DNA molecule early in 1953, described it as "just a beautiful helix," and in the first draft of the famous DNA discovery paper he wrote with Francis Crick he referred to the "very beautiful" work of Franklin and the other scientists at King's College. At the insistence of colleagues, though, the phrase was cut out of the final version. In spontaneous and unguarded moments, scientists apply the word "beautiful" to results, techniques, instruments, equations, theories—and, perhaps most intriguingly, to the engines of scientific advance: experiments.[2]

When scientists speak of beauty in these contexts they generally employ the word loosely, equivocally, and sometimes even in a contradictory way. One can hardly blame them; is there any subject more difficult to discuss with precision? Victor Weisskopf, one of the great twentieth-century physicists, remarked in 1980 that "what's beautiful in science is the same thing that's beautiful in Beethoven." Yet just a few years later he wrote that "what is usually referred to as 'beauty' in science has little to do with beauty as we experience it in art."[3] Weisskopf sensed both a similarity and a difference between beauty in science and art, but had no way to articulate this difference coherently.

Other scientists, however, have attempted to speak about the issue carefully. One is British mathematician G. H. Hardy, who, in his marvelous book, *A Mathematician's Apology*, cites several mathematical proofs as being beautiful, then defends the claim. Hardy proposes that the essential criteria for beauty in his field are unexpectedness, inevitability, and economy—and also depth, or how fundamental a proof is. Thus a mathematical proof can be called beautiful and a chess problem cannot, Hardy says. The solution to a chess problem cannot change the rules of the game, but a novel mathematical proof can alter mathematics itself.[4]

The nineteenth-century British physicist Michael Faraday was famous for his public lectures at the Royal Institution in London. One of his most popular was "The Chemical History of the Candle." At the beginning of his talk, Faraday described candles as "beautiful." He explained that he was not referring to prettiness of color or shape; in fact, Faraday disliked gaudy ornamental candles.

Instead, he said, beauty means "not the best-looking thing, but the best-*acting* thing." In his eyes, a candle is beautiful because its functioning elegantly and efficiently rests upon a wide range of universal laws. The heat of the flame melts the wax while drawing upward currents of air to cool the wax at the edge, thus creating a cup for the molten wax. The pool of wax remains horizontal because of "the same force of gravity which holds worlds together." Capillary action draws melted wax up the wick from the "cup" at the bottom of the wick to the flame at the top, while the flame's heat triggers a chemical reaction in the wax that sustains the flame. The beauty of the candle, Faraday said, lies in the intricate play of scientific principles upon which it depends, and in the economy with which it knits them together.[5]

What about the beauty of an experiment? An experiment, unlike a painting or a sculpture, is dynamic. It is more like a dramatic performance, for it is something that people plan, stage, and observe in order to produce something they are critically interested in. How can we know the circumference of the earth without stretching a tape measure around the equator? How can we tell that the earth rotates without flying into outer space and watching, or know what's inside an atom? By carefully staging an event in the laboratory—sometimes with simple objects such as prisms and pendulums—we can make the answers appear right before our eyes. Form emerges out of chaos—and not magically, like a prestidigitator pulling a rabbit out of a hat, but because of events we ourselves orchestrate. We make the world's mysteries speak.[6]

The beauty of an experiment lies in *how* it makes its elements speak. Hardy's comparison of proof and a chess problem suggest that a beautiful experiment is one that shows something deep about the world in a way that transforms our understanding of it. Faraday's evocation of the candle's beauty suggests that an experiment's elements have to be efficiently arranged. And both Hardy and Faraday suggest that a beautiful experiment should be definitive, revealing its result without need for further generalizations or inferences. If the beautiful experiment raises questions, they are more about the world than the experiment itself.

Each of these three elements of beauty—depth, efficiency, and definitiveness—appears consistently in the more formal and systematic accounts of the beautiful that philosophers and artists have provided for centuries. Some, from Plato to Martin Heidegger, emphasize the way that a beautiful thing points beyond itself to the true and the good; it is the irruption of the one in the many, the infinite in the finite, the divine in the worldly. Others, such as Aristotle, focus more on the composition of the beautiful object, emphasizing the role of symmetry or harmony, that each of its elements contributes something essential. Finally, still others—including David Hume and Immanuel Kant—stress the particular kind of satisfaction that the beautiful object incites in us. Sometimes we may not realize what our expectations are until they are fulfilled, but the beautiful object brings with it the joyful realization, "*That's* what I really wanted!" The fact that experiments can possess these properties suggests that they can indeed be called "beautiful"—and not in a metaphorical way, by stretching the proper meaning of the term, but legitimately, in the old-fashioned, meaty sense of the word.

In *The Innocents Abroad*, Mark Twain recounts his visit to the Baptistery in the Duomo of Pisa, where he was shown the famous swinging chandelier that, according to legend, had inspired the seventeen-year-old Galileo to consult his pulse and, in a crude and improvised experiment, to discover that the pendulum's swing was isochronous; that is, it took the same time to travel back and forth regardless of the distance covered. (A pendulum's isochrony, as Twain knew, is the principle underlying most mechanical clocks.) Twain found the pendulum both patrician and proletarian; looking at it, he was filled with awe at Galileo's discovery, which allowed humankind to demarcate the hours, and experienced a newfound intimacy with the world.

> It looked like an insignificant thing to have conferred upon the world of science and mechanics such a mighty extension of their dominions as it has. Pondering, in its suggestive presence, I seemed to see a crazy universe of swinging disks, the toiling children of this sedate parent. He appeared to have an intelligent expression about him of knowing that he was not a lamp at all; that he was a Pendulum; a pendulum disguised, for prodigious and inscrutable purposes of his own deep devising, and not a common pendulum either, but the old original patriarchal Pendulum—the Abraham Pendulum of the world.[7]

In Twain's inimitable fashion, his remarks illustrate the beauty that even a rudimentary scientific experiment can have if it reveals something deep about the world, shows it in a simple and direct way, and does so in a manner that satisfies us without requiring further proof.

The swing of a chandelier, beams of light through a set of prisms, the slow procession of a pendulum's plane of oscillation around in a circle, the near-simultaneous descent of falling objects of different weights released simultaneously, the ratios of the velocities of oil droplets—all these events, when staged a certain way, can reveal something about themselves and about the world. They are at once like landscape paintings, which please, compel, and enlighten us, and like maps, which guide us more deeply into the world. An experiment is a threshold event: It may make use of ordinary and uncomplicated things, but these serve as the bridge to a domain of meaning and significance. Beauty conducts us into the world of ideas while still anchoring us in the world of sense, as the

German poet and philosopher Friedrich Schiller often insisted. "Beauty is the moment of transition, as if the form were just ready to flow into other forms," wrote the American essayist Ralph Waldo Emerson.[8]

The beauty of experiments can take many forms—just as the beauty of a piece by Bach is different from one by Stravinsky. Some have a synoptic beauty, drawing together different universal laws, while others have a beauty of breadth, linking elements on vastly different scales. Some have an austere beauty, engaging us with their stark simplicity to reveal pure form, while others are sublime, compelling us with hints of nature's ultimately incomprehensible limitlessness and terrifying power. Most beautiful experiments involve elements of each kind.

YOU MIGHT THINK of this book as a special kind of gallery. The gallery contains items of rare beauty, each with its own design, distinct materials, and unique appeal. You will not like everything equally, for your background, experience, education, and personal taste will incline you to prefer some items over others.

One of the most difficult tasks in setting up a gallery is selecting what goes into it. I coped with this problem as follows. In 2002, sparked by yet another scientist talking about a beautiful experiment, and recalling not only Glashow's remark but hundreds of others like it that I had heard over the years, I conducted a poll. I asked readers of the international magazine *Physics World*, for which I am a columnist, what they thought were the most beautiful experiments. To my surprise, my readers sent in more than three hundred candidates. These ranged from actual historical experiments to thought experiments, proposed experiments, proofs, theorems, and models. They ranged over every scientific field, from physics to psychology. My poll was picked up by Weblogs and Internet discussion groups, which provided me with hundreds of other candidates. To compile my list of the most beautiful scientific experiments, I selected the ten most frequently mentioned candidates.[9] Some people will object that this list is dominated by physics experiments, and it is true that my original column in *Physics World* asked readers to name the most beautiful physics experiments. Still, I feel justified in claiming that this gallery of historical portraits contains the ten most beautiful science experiments. I do so mainly because the vast majority of my respondents, from *Physics World* readers and elsewhere, did indeed interpret my survey to mean science experiments, and because even the suggestions from *Physics World* readers ranged into chemistry, engineering, and physiology.

Also, over half the experiments on the list were first carried out before physics became a separate branch of science. Finally, they are classic textbook exercises, frequently discussed and performed when historical experiments are taught in basic science education courses and have become emblematic of science in the broadest sense. It is no surprise, therefore, that allusions and descriptions of these dramatic and epochal experiments appear in the works of artists as diverse as playwright Tom Stoppard, musician Philip Glass, and novelist Umberto Eco, and frequently crop up in popular culture.[10]

I have chosen to arrange these experiments in chronological order. Doing so yields a powerful sense of the vastness of the journey science has taken over almost 2,500 years. This list takes us from a time when the pressing issues of science included obtaining the roughest of estimates of the basic properties of the earth—among these its size and position in the heavens—to the era in which scientists began making precise measurements of the properties of the atom and its constituent particles. It takes us from the time of simple, homemade tools like sundials and inclined planes to the time of advanced instrumentation. It takes us from the time when scientists worked alone (or at most with an assistant or two) to the present, when scientists often work in teams of hundreds. It provides a glimpse into the personalities and the creative thinking of some of the field's most interesting figures. Many landmark experiments in the evolution of science appear here: Galileo's experiment with inclined planes established for the first time a mathematical formula for accelerated motion; Isaac Newton's *experimentum crucis* unraveled the nature of light and colors; Thomas Young's two-slit experiment revealed the wavelike character of light; and Ernest Rutherford's discovery of the atomic nucleus inaugurated the nuclear age. This list contains experiments that either powerfully illustrate or helped motivate some of the great paradigm shifts in science, from the Aristotelian to the Galilean perspective on motion, from the corpuscular to the wave picture of light, and from classical to quantum mechanics.

With one exception, these experiments were preferred by about the same number of people: Thus I do not rank them. That exception, the two-slit experiment illustrating the quantum interference of single electrons, was far and away the most frequently mentioned candidate for the most beautiful science experiment. Critics will inevitably bicker about my choices. But they will be arguing about the selection process and not with the gallery's theme: the beauty of scientific experiments.

The Prism and the Pendulum

Earliest known datable hour-counter, from the third century B.C., during Eratosthenes' lifetime. While mostly intact, the gnomon, or pointer, which cast shadows across the bowl, is missing.

One

Eratosthenes' Measurement of the Earth's Circumference

—

IN THE THIRD CENTURY B.C., A GREEK SCHOLAR NAMED Eratosthenes (ca. 276–ca. 195 B.C.) made the first known measurement of the size of the earth. His tools were simple: the shadow cast by the pointer of a sundial, plus a set of measurements and assumptions. But the measurement was so ingenious that it would be cited authoritatively for hundreds of years. It is so simple and instructive that it is reenacted annually, almost 2,500 years later, by schoolchildren all around the globe. And the principle is so elegant that grasping it all but makes you want to go measure the length of a shadow.

Eratosthenes' experiment combined two ideas of far-reaching importance. The first was to picture the cosmos as a set of objects (the earth, sun, planets, and stars) in ordinary three-dimensional space. This may seem obvious to us but was not widely believed then; it was a Greek contribution to science to insist that underneath the myriad, ever-changing motions in the world and in the night sky lay an impersonal and changeless order, a cosmic architecture that could be described and explained by geometry. The second idea was to apply ordinary measurement practices to understand the scope and dimensions of this cosmic architecture. In combining these two ideas, Eratosthenes came up with the audacious notion that the same techniques that had been developed for building houses and bridges, laying out fields and roads, and predicting floods and monsoons could provide information about the dimensions of the earth and other heavenly bodies.

Eratosthenes began by assuming that the earth was approximately round. For despite the widespread contemporary belief that Columbus set out to prove that the world was not flat, many of the ancient Greeks who had thought carefully about the cosmos had already concluded that the earth not only had to be round, but also that it must be relatively tiny in comparison with the rest of the universe. Among these scholars was Aristotle, whose book *On the Heavens,* written about a century before Eratosthenes, advanced several different arguments, some logical and some empirical, for why the earth had to be spherical. Aristotle pointed out, for instance, that during eclipses the shadow cast by the earth on the moon is always curved—something that would occur only if the earth were round. He also noted that travelers see different stars when going north or south (unlikely if the world is flat), that certain stars visible in Egypt and Cyprus cannot be seen in more northern lands, and that certain other stars always visible in the north rise and set in the south, as if they were seen at a distance from the surface of a round object. "This indicates not only that the earth's mass is spherical in shape," Aristotle wrote, "but also that as compared with the stars it is not of great size."[1]

But the resourceful thinker also offered more creative arguments. From accounts by foreign travelers and military expeditions, he knew that elephants were found in distant lands both to the east (Africa) and to the west (Asia). Therefore, he said, these lands are probably joined—a clever if incorrect guess. Other Greek scholars made additional arguments for the earth's spherical shape, including the difference in the time of sunrise and sunset in different countries, and the way departing ships gradually fall out of sight from the hull upward.

None of this, though, answered a basic question: How big is this round earth? Was it even possible to know its size, without having surveyors actually pace over its entire circumference?

Until Eratosthenes, we know of only estimates of the earth's size. The earliest is by Aristotle, who wrote that "those mathematicians who try to calculate the size of the Earth's circumference arrive at the figure 400,000 stades," but he did not tell us either his sources or their reasoning.[2] It is also impossible to convert his figure precisely to modern units. A stade, or "stadium," referred to the length of a Greek racetrack, which varied from city to city. By using a rough estimate for the stade, today's researchers have put Aristotle's figure at somewhat more than 40,000 miles (the actual number is about 24,900 miles). Archimedes, who built models of the cosmos in which the heavenly bodies revolved around one another, gave a slightly smaller estimate than Aristotle's: 300,000 stades, or more than 30,000 miles. But he, too, provided no clue to his source or reasoning.

Enter Eratosthenes. A younger contemporary of Archimedes, Eratosthenes was born in North Africa and educated in Athens. He was a polymath, expert in many areas from literary criticism and poetry to geography and mathematics. But he was not thought to have achieved the first rank in any, which led associates to give him the sarcastic nickname "Beta," the second letter of the Greek alphabet, in a joke meaning he was always second best. Despite the gibes, his brilliance was so renowned that in the middle of the third century B.C. the king of Egypt invited Eratosthenes to tutor his son, and later appointed him to head the famous library at Alexandria. This was the first and largest library of its kind, and had been established by the Ptolemys, the kings of Egypt, in the course of building up Alexandria as the cultural capital of the Greek world. The library became a meeting point for scholars around the world, and Alexandria grew into an important intellectual crossroads—it was home to Euclid and his school, for instance. At Alexandria, the librarians were able to amass a comprehensive collection of manuscripts on a wide range of subjects that anyone with the appropriate scholarly credentials could use. (The Alexandrian library was also the first one known to have arranged manuscripts by author in alphabetical order.)

Eratosthenes wrote two books on geography that were of particular importance in the ancient world. *Geographica,* a three-volume set, was the first to map the world using parallels (lines parallel to the equator) and meridians (longitudinal lines, which pass through both poles and a given location). His *Measurement of the World* contained the first known description of a way to measure the size of the earth. Unfortunately, both works have been lost, and we have to reconstruct Eratosthenes' reasoning from the remarks of other ancient authors who knew them.[3] Fortunately, many did.

Eratosthenes began by reasoning that if the earth were a small, spherical body in a vast universe, then other parts, such as the sun, must be far away—so far away that its rays were essentially parallel no matter where they struck the earth. He also knew that as the sun crept upward in the sky, shadows grew progressively shorter—and he knew from travelers' reports that on the summer solstice in the town of Syene (modern-day Aswan), the sun stood directly overhead, and shadows would disappear around anything vertical, including columns, poles, and even gnomons, the vertical indicators or pointers in sundials, whose sole function was to cast shadows. Shadows even seemed to disappear from the town well as the light flooded it uniformly, "like a stopper fitting evenly all round to the mouth," according to one ancient source.[4] (I exaggerate a little; the shadows did not disappear completely, but simply fell straight below objects, rather than off to the side as they usually do.)

Furthermore, Eratosthenes knew that Alexandria was north of Syene and roughly on the same meridian. And thanks to the royal surveyors, whom the Egyptian government sent to pace and remap the land each year after the seasonal flooding of the Nile, he knew that the two towns were about five thousand stades apart (the number was rounded off, so we cannot use this information to establish a precise equivalent between stades and modern units).

In modern terms, Syene was on the Tropic of Cancer, an imaginary line around the world that passes through northern Mexico, southern Egypt, India, and southern China (it is shown on most globes). All the points on the tropic share a single unusual feature: The sun is directly overhead only once a year, on the longest day of the year—June 21, the summer solstice. People who live north of the Tropic of Cancer never see the sun directly overhead, and the sun always casts a shadow. People in the northern hemisphere who live south of it see the sun directly overhead twice a year, once before the solstice and once after, with the exact day depending on the location.

The reason for this has to do with the position of the earth, whose axis is tilted with respect to the sun. But this did not concern Eratosthenes. What mattered to him was that when the sun was directly overhead at Syene, it would not be overhead anywhere to the north or south—including Alexandria—and a gnomon would cast a shadow in those places. How long a shadow would depend on the amount of the earth's curvature; if the curvature were great, a shadow in a place like Alexandria would be longer than if the earth's curvature were slight.

Thanks to his geometrical knowledge, Eratosthenes knew enough to devise an ingenious experiment that would tell him the exact amount of curvature, and so the circumference of the earth.

To appreciate the beauty of this experiment, we do not need to know anything about the specific way Eratosthenes went about it. This is lucky, for we do not even have his description of what he did. We know it only through incomplete secondhand descriptions by his contemporaries and successors, most of whom clearly did not understand all its details. We do not need to know anything about his path of inquiry—what specifically motivated his interest in this problem, what his initial steps were, what backtracks, if any, he made, how he achieved the final realization, and in what further directions he was led. This is unfortunate, for it may create the impression that the idea came to him as a brainstorm, a bolt from the blue, but it does not impede our ability to understand the experiment. Nor are we required to engage in speculative intellectual leaps, to follow complex mathematical reasoning, or to employ clever empirical guesses based on things like elephant demographics. The beauty of his experiment lies in the way it makes possible the discovery of a dimension of cosmic proportions by measuring the length of a tiny shadow.

Its breathtaking simplicity and elegance can be captured in two diagrams, Figures 1.1 and 1.2.

During the solstice, when the sun is directly overhead at Syene (A), shadows disappeared—they fell straight toward the center of the earth (line AB). Meanwhile, the shadows at Alexandria (E) also fell in the same direction (CD), because the sun's rays are parallel; but because the earth is curved they fell at a slight angle, which we'll call x. A small angle or short shadow would mean that the earth's curvature was relatively flat and that the earth had a large circumference; a large angle or long shadow would mean a sharp curvature and a small circumference. Was there a way to figure out the circumference exactly by the length of the shadow? Geometry provided a way.

According to Euclid, the interior angles of a line that intersects

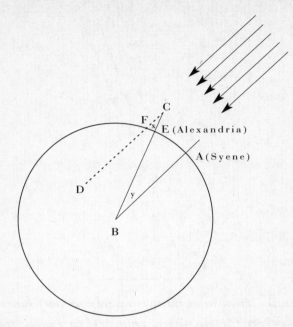

Figure 1.1.   The angle cast by shadows in Alexandria (x) is equal to the angle (y) created by the two rays that pass through Alexandria and Syene and meet at the earth's center (not to scale). Thus the fraction that the arc of a shadow (EF) in Alexandria is of a complete circle is the same as the fraction that the distance (AE) from Syene to Alexandria is of the earth's circumference.

two parallel lines are equal. Thus the angle (x) cast by shadows in Alexandria is equal to the angle (y) created by the two rays whose vertex is at the center of the earth and that pass through Alexandria and Syene (BC and BA). This in turn means that the ratio between the length of the arc of a gnomon (FE) and of the complete circle around the gnomon (see Figure 1.2) is the same as the ratio between the distance from Syene to Alexandria (AE) and the earth's circumference. If one measured this fraction, Eratosthenes realized, one could calculate the earth's circumference.

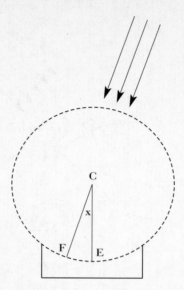

Figure 1.2. Eratosthenes could have measured either what fraction the shadow length (EF) was of the circumference of the circle described by the hour-counter's bowl, or what fraction the shadow's angle (x) was of a complete circle.

Although Eratosthenes might have made his measurement in a number of ways, historians of science are fairly sure he did it with an hour-counter, the Greek version of a sundial, because the arc of its shadow would be sharply defined. An hour-counter, or *skaphe,* consisted of a bronze bowl equipped with a gnomon, whose shadow crept slowly across hour-lines cut into the bowl's surface. But Eratosthenes used this piece of equipment in a new way. He was not interested in the position of the shadow on the hour-lines to track the passage of time, but rather in the angle of the shadow cast by the gnomon at noon on the summer solstice. He would measure what fraction that angle was of a complete circle (the practice of measuring in degrees obtained by dividing a circle into 360 equal parts would not become standard for about another century). Or, in what amounts to the same thing, he might have measured the ratio of the length of the arc cast by the gnomon on the bowl to the full circumference of the bowl.

At noon on that day, Eratosthenes ascertained that the shadow's wedge was 1/50th of a complete circle (we would say 7.2 degrees). The distance between Alexandria and Syene was therefore a fiftieth of the distance around the entire meridian. Multiplying 5,000 stades by 50, he got 250,000 stades as the earth's circumference; later, he adjusted this to 252,000 stades (both work out to just over 25,000 miles). The reason for this adjustment is not clear, but probably had to do with his desire to simplify the calculation of geographical distances. For Eratosthenes was in the habit of dividing circles into sixty parts, and a circumference of 252,000 stades made for an even 4,200 stades per each sixtieth part of the circle. But whether one uses 250,000 or 252,000 stades, and whichever figure one trusts for converting his stades into modern units of distance, his estimate lies within just a few percent of the figure in use today of 24,900 miles.

Eratosthenes' picture of the cosmos was critical to the success of the experiment. Without this particular picture, measuring the shadow would not yield the earth's circumference. For example, an ancient Chinese cartographic text, the *Huainanzi*, or "Book of the Master of Huainan," notes that gnomons of the same height but at different (north-south) distances from one another cast shadows of different lengths at the same time.[5] On the assumption that the earth is essentially flat, the author attributed this difference to the fact that the gnomon casting the narrower shadow is more directly under the sun, and argues that the difference in shadow lengths can be used to calculate the height of the sky!

Eratosthenes' data, and his measurement, were approximate. He was probably aware that Syene is not precisely on the Tropic of Cancer. Nor does it lie exactly due south of Alexandria. The distance between the two towns is not exactly five thousand stades. And because the sun is not a point of light but a little disc (about half a degree wide), the light from one side of the disc does not strike the gnomon at exactly the same angle as the light on the other side, smearing the shadow slightly.

But given the technology Eratosthenes had at his disposal, the experiment was good enough. His figure of 252,000 stades was accepted by the ancient Greeks as a reliable value for the earth's circumference for hundreds of years. In the first century A.D. the Roman author Pliny praised Eratosthenes as an "outstanding authority" on the circumference of the earth, his experiment as "audacious," his reasoning as "subtle," and his figure as "universally accepted."[6] About a century after Eratosthenes, another Greek scholar tried using the difference between the angle at which the bright star Canopus was visible from Alexandria and the angle of the same star seen from Rhodes (where the star was said to be right on the horizon) to measure the earth's circumference, but the result was unreliable. Even a millennium afterward, Arabian astronomers were unable to improve on his work, though they tried by such means as measuring the earth's horizon as seen from a mountaintop of known height and measuring the height of a star from the horizon from two different locations simultaneously. Eratosthenes' measurement would not be bettered until modern times, when much more exact measurements of the positions of celestial bodies became possible.

The experiment transformed geography and astronomy. First, it allowed any geographer to establish the distance between any two locations of known latitude—between Athens and Carthage, for instance, or between Carthage and the mouth of the Nile. It allowed Eratosthenes to work out the size and position of the known inhabited world. And it provided Eratosthenes' successors with a yardstick for trying to determine other cosmic dimensions, such as the distances to the moon, sun, and stars. In short, Eratosthenes' experiment transformed the picture human beings had of the earth, of the earth's position in the universe (or at least the solar system), and of the human place in it all.

Eratosthenes' experiment, like any other kind of performance, is abstract in the sense that it does not depend on any specific realization and can be performed in many ways. It was, therefore, a contribution to human culture. The ingredients are ordinary and familiar: a shadow, a measuring tool, junior high school geometry. One does not need to be in Alexandria or to use a *skaphe;* one does not even need to do it during the solstice. Hundreds of schools around the globe have Eratosthenes' experiment on their curriculum. Some use shadows cast by makeshift sundials, others by flagpoles or towers. Often these reenactments are done in collaboration with other schools via e-mail, using a geography Web site to determine latitudes and longitudes and MapQuest to determine distance. These reenactments of Eratosthenes' experiments are not like replications of say, the Battle of Gettysburg by Civil War enthusiasts, where the aim is historical accuracy or at least an entertaining simulation. The students are not copying or simulating Eratosthenes' experiment—they are actually doing it, as if for the first time, and the experiment shows the result freshly, right before their eyes, and so directly that it can scarcely be doubted.

Eratosthenes' experiment also dramatically illustrates the nature of experimentation itself. How can scientists possibly know something like the circumference of the earth without actually physically measuring it? We are not helpless, nor do we have to wait for brute-force methods like measuring tapes tens of thousands of miles long. A cleverly staged performance, using the right props, can trick even ephemeral and fluid things like shadows into disclosing the fixed and unchanging dimensions of the heavens. Eratosthenes' experiment shows the way we can establish form out of chaos, or even out of fleeting shadows, by devices of our own making.

The beauty of Eratosthenes' experiment arises from its awesome breadth. Some experiments draw order out of chaos by the way they analyze, isolate, or dissect something before us. This experiment points our attention in the opposite direction, and measures vastness in small things. It expands our perception, providing us with new ways of looking at the seemingly simple question, "What are shadows, and how are they formed?" The experiment makes us realize that the dimension of *this* particular transient shadow is connected with the roundness of the earth, with the size and remoteness of the sun, with the constantly changing positions of these two bodies, and with all the rest of the shadows on the planet. The sun's vast distance from us, time's cyclical progression, and the earth's roundness acquire an almost palpable presence in this experiment. It thus affects the quality of our experience of the world.

Experiments in the physical sciences are often thought to be impersonal, and to diminish humanity's significance in the universe. Science is thought to strip humanity of its privileged position—and some people compensate for this imagined loss by engaging in magical thinking, fantasizing that the sun, planets, and stars have a mystical tie to their personal destinies. But Eratosthenes' seemingly abstract experiment humanizes us in a more genuine way by giving us a realistic sense of who and where we are. While most everything around us celebrates bigness, immediacy, and domination, this experiment fosters an appreciation for the disclosive power of smallness, of temporality, and of the way that things of all dimensions are ultimately interconnected.

WHY SCIENCE IS BEAUTIFUL

DO WE *HAVE* TO CALL Eratosthenes' experiment beautiful? Even if it satisfies the three criteria I mentioned in the introduction—if it shows us something fundamental, efficiently, and in a way that satisfies us and leaves us with questions not about the experiment but about the world—one can still object to calling it beautiful. I have heard people protest, for instance, that to speak about the beauty of experiments is irrelevant, elitist, and seductive.

Those who say the beauty of experiments is irrelevant generally mean that beauty is the realm of subjectivity, opinion, and emotion, while science is the realm of objectivity, fact, and intellect. Some may say, for instance, that to call experiments "beautiful" confuses what the arts and humanities do (i.e., explore and expand human life and culture) with what the sciences do (i.e., describe the natural world). Or they may say it commits what philosopher Benedetto Croce called "the intellectualist error," illegitimately mixing art and ideas. Painter and critic John Ruskin builds a respect for this division into his very definition of beauty: "Any material object which can give us pleasure in the simple contemplation of its outward qualities without any direct and definite exertion of the intellect, I call in some way, or in some degree, beautiful."[1] We do not like to have to think to appreciate our beauties. Because science experiments are creatures of the intellect, this objection runs, they do not belong on lists of beautiful things.

Those who say the beauty of experiments is elitist take this objection further. Beauty, they point out, can only be intuited and must be grasped firsthand—imagine trying to appreciate the beauty of a van Gogh painting or a Mozart concerto by reading a description of it. The beauty of a science experiment, therefore, can be apparent only to scientists. J. Robert Oppenheimer once remarked that, for outsiders, trying to understand the birth of quantum mechanics— which he called a time of "terror as well as exaltation"—would be like having to listen "to accounts of soldiers returning from a campaign of unparalleled hardship and heroism, or of explorers from the high Himalayas, or of tales of deep illness, or of a mystic's communion with his God," adding that "such stories tell little of what the teller has to tell." The beauties in that world—and there are said to be many—are only accessible to its inhabitants. A large section of the mansion of beauty, evidently, is off-limits to nonscientists. But this is anathema to modern democratic sensibilities, and smacks of elitism.

A third and most forceful objection is the seduction argument. Scientists may say that their job is to find theories that work, and that it is distracting at best, and dangerous at worst, if practitioners become self-conscious about creating beautiful objects.[2] Practitioners may shackle their intellects and "go soft" paying attention to beauty—only the aesthetically ruthless are truly ready for the imaginative and insightful work of science. Nonscientists, meanwhile, may fear that speaking of beauty in science is not only superficial and sentimentalizing, but that it disguises a hidden public-relations agenda. It is easy to sympathize. The images that accompany most talk about beauty in science that I have come across originated not at the workbench but in public-relations departments. At one talk I attended, the final slide was the famous picture of the earth rising above the moon's surface. That picture is indeed beautiful. But though it has admirably served NASA's publicity juggernaut for decades, astronomers have never used this picture as data.

These three objections are based on misunderstandings of beauty. The first one mistakes beauty for ornament. To aestheticize science—to look at its external appearances—is the quickest way to lose sight of its beauty. An experiment's beauty lies in *how* it shows what it does. As we shall see, the beauty of Newton's *experimentum crucis* has nothing to do with the colors produced by his prisms (in fact, he had to look past the colors to come up with the experiment) but with how it reveals what it does about light. The beauty of Cavendish's experiment to weigh the world has nothing to do with the outward appearance of his monstrosity of an instrument but with its austere precision. And the beauty of Young's experiment is not due to its rather mundane pattern of black-and-white stripes but to how these disclose something essential about light.

The second objection, like the first, fails to appreciate how intrinsically bound up our (educated) perception is with feelings and emotions. While we aren't naifs in the laboratory, we aren't naifs in the art museum, either. We exercise educated perception in apprehending the beauty of painting, music, and poetry, and we can also easily *fail* to recognize the beauty of things that require little "exertion of the intellect" to grasp. (For example one of Pablo Neruda's poems, "Ode to My Socks," describes the beauty of pieces of footwear.) The effort needed to grasp the beauty of experiments— and appreciating the beauty of the ten experiments in this book does not require much—is not an obstacle. The real obstacle may be our tendency to view everything around us instrumentally, in terms of how they serve our goals. Our appreciation of beauty thus may simply be slumbering and need rousing. And, as Willa Cather wrote, "Beauty is not so plentiful that we can afford to object to stepping back a dozen paces to catch it."[3]

The third objection is the strongest and deepest. It is a version of the old conflict between reason and art that was already ancient in Plato's time—the fear that human beings are more readily enraptured by appearances rather than convinced by logic. For Plato, in the *Republic*, the arts cater to the passions rather than to reason, "gratifying the soul's foolish part" and leading us astray.[4] St. Augustine was another who saw danger in the ability of the senses to overwhelm reason, and warned of the danger posed even by church music, confessing that he sometimes finds "the singing itself more moving than the truth which it conveys." This, he continues, "is a grievous sin, and at those times I would prefer not to hear the singer."[5] This third objection amounts to a scare story: Beware the magical and seductive power of images; stick to reason and logic. Many logic-oriented philosophies thus divorce, or even oppose, truth and beauty. "The question of truth," wrote logician Gottlob Frege in one of his most influential works, "would cause us to abandon aesthetic delight for an attitude of scientific investigation."[6]

The answer to this third objection brings us to the heart of science, and of art. It requires appealing to philosophical traditions other than those dominated by models of logic and mathematics. These traditions appeal to a more fundamental view of truth as the disclosure of something rather than its accurate representation (as Heidegger insistently points out, *aletheia*, the Greek word for truth, means literally "unconcealing"). Such traditions open the way to see scientific inquiry as integrally linked with beauty. Beauty is not a magic power beside and apart from the disclosure of truth, but accompanies it: an unconscious by-product of science, so to speak. Beauty is the talisman for achieving a new foothold on reality, unshackling our intellects, deepening our engagement with nature. To this extent beauty is to be contrasted with elegance, in which that new foothold is lacking.[7] "Beauty" describes a tuning or fit between an object that discloses a new foothold, and our openness to what is disclosed.[8]

Does Eratosthenes' experiment really do this?

It is indeed possible to see this experiment abstractly, as a third-century B.C. version of a global positioning system, as a problem of quantification or an intellectual exercise. This is how most of my classmates saw it when we were taught it in junior high school, and how our teacher presented it. But to see it this way we first have to suffocate our imaginations—abetted by the overriding desire to get the right answer, by conventional science instruction, and by our exposure to satellite photographs. Most of the time we ignore shadows, those epiphenomena of light, or think, "How nice!" and move on. But Eratosthenes' experiment shows us that each shadow on the sunlit earth is woven together with all the others in a continuously evolving whole. Contemplating Eratosthenes' experiment stimulates rather than strangles our imagination, driving us out of our routines and making us stop and become more aware of where we are in the universe. This experiment reawakens our wonder at the world.

If we are serious about beauty, then yes, Eratosthenes' experiment *is* beautiful. Like other beautiful things, it simultaneously sets the world at a distance to linger over, and thrusts us into it more deeply.

The Leaning Tower of Pisa

Two

DROPPING THE BALL:

The Legend of the Leaning Tower

—

COMMANDER DAVID R. SCOTT: WELL, IN MY LEFT HAND, I have a feather; in my right hand, a hammer. And I guess one of the reasons we got here today was because of a gentleman named Galileo, a long time ago, who made a rather significant discovery about falling objects in gravity fields. And we thought: Where would be a better place to confirm his findings than on the moon?

[Camera zooms in on Scott's hands, one of which holds a feather and the other a hammer, then pulls back to take in the entire background, including the Apollo 15 landing craft known as the Falcon.*]*

SCOTT: And so we thought we'd try it here for you. The feather happens to be, appropriately, a falcon feather for our *Falcon.* And I'll drop the two of them here and, hopefully, they'll hit the ground at the same time.

[Scott releases hammer and feather, which fall side by side, and just over a second later they hit the lunar surface more or less simultaneously.]

SCOTT: How about that! Mr. Galileo was correct in his findings.[1]

ACCORDING TO LEGEND, the experiment at the Leaning Tower of Pisa established convincingly for the first time that objects of different weights fall at the same rate, thereby overthrowing the authority of Aristotle. This legend is associated with a single person (the Italian mathematician, physicist, and astronomer Galileo Galilei), with a single place (the Leaning Tower of Pisa), and with a single episode. How much truth is in this legend, and what are its remaining mysteries?

Galileo (1564–1642) was born in Pisa to a musical family. His father, Vincenzio, was a well-known lutist with a penchant for controversial experiments, conducting investigations into intonations, intervals, and tuning in a way that championed the evidence of the ear against the authority of ancient scholars. Vincenzio's son was equally strong willed. Galileo's biographer Stillman Drake once picked out two features of his personality that were essential to his scientific success. The first was Galileo's "pugnacious disposition," which made him unafraid and even eager to engage in battles "to overthrow tradition and vindicate his scientific position." The second was that Galileo's personality was balanced between two extremes of temperament. The one "delights in observing things, notes resemblances and relationships among them, and forms generalizations without being unduly disturbed by apparent exceptions or anomalies"; the other "frets and worries over any unexplained deviation from a rule [and] may even prefer no rule at all over one that does not always work with mathematical precision." Both traits are valuable in science, and all scientists possess some mixture of each, though usually one of them predominates. But Galileo's temperament, Drake says, was evenly balanced between these two extremes.[2] Also essential to Galileo's impact on the world was the literary skill with which he was able to address and persuade those around him.

Galileo entered the University of Pisa probably in fall 1580 intending to study medicine, but became fascinated by mathematics. He landed a position at the university as a lecturer in 1589, and began to investigate the motion of falling bodies. He remained at the University of Pisa for three years; if the Leaning Tower experiment ever took place, it would have been during this period. In 1592 Galileo moved to Padua, where he lived for eighteen years and did most of his important scientific work, including the construction of an astronomical telescope. This telescope allowed him to make astronomical discoveries; Galileo was the first to observe the moons of Jupiter, for example. And it also provided the first opportunity for Galileo to stir controversy, since his astronomical discoveries contradicted the Ptolemaic system (in which the sun moves around the earth) as well as the Aristotelian account of motion, and supported the Copernican system (in which the earth moves around the sun). In Padua, too, he became famous for his elaborate demonstrations of physical laws, delivering his lectures in a hall built for two thousand people. In 1610, he moved to Florence to the court of the Grand Duke of Tuscany. In 1616, Galileo was warned not to "hold or defend" the Copernican doctrine, but sixteen years later, in 1632, he published a brilliant book, *Dialogue Concerning the Two Chief World Systems—Ptolemaic and Copernican*, which, though approved by the censors, was recognized as building a strong case for the Copernican system. The following year, 1633, Galileo was summoned to Rome by the Catholic Church and forced to say that he "abjured, cursed, and detested" his erroneous views. He was sentenced to what amounted to house arrest, spending his last years in a town called Arcetri, outside of Florence. Shortly before his death, Galileo acquired the services of a promising young mathematician named Vincenzio Viviani, who became the disciple and faithful secretary of the now-blind scientist and listened patiently to his recollections, ruminations, and tirades. Dedicating himself to preserving Galileo's memory, Viviani eventually wrote Galileo's first biography.

We owe many famous Galilean legends to Viviani's warm biography. One is the story of the Abraham Pendulum: how Galileo, while still a medical student in 1581, used his own pulse to time the swing of a chandelier hanging in the Baptistery in the Duomo of Pisa and discovered its isochrony. Historians know the story cannot be entirely accurate: the chandelier that hangs there today was installed in 1587. But there may be a grain of truth to the story, because its precursor surely obeyed the same laws of physics. The most famous Viviani story recounts how Galileo climbed to the top of the Leaning Tower of Pisa, and "in the presence of other teachers and philosophers and all the students," and by "repeated experiments," showed that "the velocity of moving bodies of the same composition, but of different weights, moving through the same medium, do not attain the proportion of their weight as Aristotle decreed, but instead move with the same velocity."[3]

In his own books, Galileo advances arguments of various types, using logic, thought experiments, and analogies, to describe why two objects of unequal weight will fall at the same speed in a vacuum. Without mentioning the Leaning Tower explicitly, Galileo reports having "made the test" in the open air with a cannonball and a musket ball, finding that as a general rule they land at almost the same time. His fastidious mention of this deviation from what looked like the appropriate generalization as well as Viviani's failure to mention it—plus the fact that Viviani's account is our only source for the Leaning Tower episode—make many historians of science skeptical that it took place.

Whether or not Galileo actually performed the experiment at the Leaning Tower, much more was involved in his shift in thought from the Aristotelian framework to his later analyses of motion. Aristotelian natural philosophy—which included his account of motion, and is what we would call his physics—provided a coherent and fully articulated system based on the idea of a stationary central earth and a heavenly realm in which objects behave very differently from the way they do on earth. For Galileo to doubt, and then to challenge, the Aristotelian system meant doubting and challenging both these aspects—Aristotle's idea of a stationary earth as well as his account of earthly motion.

A central feature of the Aristotelian view of the universe was that heaven and earth were distinct realms made up of different kinds of substances and governed by separate laws. The motions up in the heavens were orderly, precise, regular, and mathematical, while the motions down here on earth were messy and irregular, and could be described only qualitatively. Moreover, the motion of bodies on earth was governed by their tendency to seek their "natural place"; for solid objects, this was downward toward the center of the earth. Aristotle thus distinguished between an upward unnatural or "violent motion" of heavy objects and their downward "natural motion."

Aristotle had observed the motion of falling bodies and noticed that their speeds seem to vary in different media depending on whether these media are "thinner" like air or "thicker" like liquids. He remarked that bodies reach a set speed as they drop, and that this speed was proportional to their weight. These ideas fit our everyday experience. If we drop a golf ball and a Ping-Pong ball from a window, the golf ball will fall faster and strike the ground first. If we drop the golf ball in a swimming pool, it falls to the bottom more slowly than in air, and is beaten to the bottom by a steel ball. Similarly, hammers fall faster than feathers. Aristotle had codified this in a framework—what philosophers of science would later call a "paradigm"—oriented by the everyday phenomena he was seeking to explain. For instance, an agent (such as a horse) faced obstacles (friction and other types of resistance) keeping a body (a cart) in motion. In these familiar circumstances, motion almost always represents a balance between force and resistance. Aristotle therefore approached the case of falling bodies as one in which a force (a natural tendency, as he saw it, to move toward the center of the universe) was balanced by a resistance (the thickness or thinness—we would say "viscosity"—of the medium in which they are moving). He also concluded that, in the absence of a resisting medium, the speed of falling bodies would be infinite.

In modern terms, Aristotle's approach fails to incorporate acceleration adequately. Scholars had begun to suspect something like this long before Galileo. As early as the sixth century A.D. the Byzantine scholar Joannes Philoponus wrote of experiments contradicting Aristotle: "For if you let fall from the same height two weights of which one is many times as heavy as the other, you will see that the ratio of the times required for the motion does not depend on the ratio of the weights, but that the difference in time is a very small one." In fact, Philoponus continued, if one body weighed only twice as much as the other, "there will be no difference, or else an imperceptible difference, in time."[4]

In 1586, before Galileo went to Padua, his contemporary, the Flemish engineer Simon Stevin, wrote of experiments showing that Aristotle's account was wrong. Stevin dropped two lead balls, one of which was ten times heavier than the other, from a height of thirty feet onto a board, so that when the balls landed they would make an audible sound. "Then it will be found," Stevin wrote, "that the lighter will not be ten times longer on its way than the heavier, but that they fall together onto the board so simultaneously that their two sounds seem to be one and the same rap."[5] Aristotle, in short, was wrong on this point.

During Galileo's lifetime, several Italian scholars of the sixteenth century also wrote of experiments involving falling bodies with results that contradicted Aristotle, including a professor at Pisa (who taught there while Galileo was a student) named Girolamo Borro. Borro wrote about how he had repeatedly "tossed"—the verb he used is ambiguous—objects of equal weight but different sizes and densities, and finding each time, curiously enough, that the denser weights dropped more slowly than the others.[6]

Like the work of any great scientist whose interests were broad in scope, Aristotle's research was known to be dotted with mistakes and flaws. But until Galileo most European thinkers did not regard these blemishes as of serious import. Galileo's great achievement was to demonstrate that Aristotle's account of motion was inextricably linked with an entire scientific framework involving much more than falling bodies, and that an account of motion that adequately explained the behavior of falling bodies needed to incorporate the phenomenon of acceleration, requiring the construction of an entirely new framework. Aristotle knew that bodies picked up speed (accelerated) when dropped, but thought it not essential to free fall, rather an accidental and unimportant feature of motion that took place between the time a body was released and the time it took to reach its natural uniform speed. Galileo began by sharing this view. However, he came to realize not only the importance of acceleration, but also that it could not be simply "added" to Aristotle's system. If Aristotle were wrong about how bodies fell, his work could not be patched up but had to be completely revamped.

Galileo did not arrive at this insight immediately, though, and began with the then-normal assumption that Aristotle was correct.

And no single piece of evidence was decisive in changing his mind. Rather, he reached his revolutionary trajectory through the sum of his inquiries—the astronomical ones as well as the more mundane ones involving pendulums and falling bodies.

In his first discussion of the behavior of falling bodies, an unpublished manuscript entitled *On Motion* (written while he was at the University of Pisa), Galileo holds to Aristotle's notion that bodies fall with a uniform speed that depends on their density—one of the "general rules governing the ratio of the speeds of [natural] motion of bodies," as he puts it. A ball made of gold should fall twice as fast as one of equal size made of silver, for the first is almost twice the density of the second. Galileo evidently set out to see this actually happen—but to his astonishment and dismay was frustrated that his experiment did not work. "[I]f one takes two different bodies," he reported, "which have properties such that the first should fall twice as fast as the second, and if one lets them fall from a tower, the first will not reach the ground appreciably faster or twice as fast."[7] Historians of science have concluded from this that, even early in his career, Galileo was committed to testing theory against observation. But in the same book Galileo also makes the bizarre assertion that the lighter body at first moves ahead of the heavier body, though the heavier one eventually catches up. This has led some to doubt either Galileo's sincerity or his ability as an experimenter.

Within a few years Galileo had changed his mind about falling bodies, and had entirely abandoned the Aristotelian framework. The process of inquiry that led him to do so was complex and involved many forms of evidence and thought, not just earthly motions. Much has been reconstructed by Galileo scholars through painstaking, page-by-page analysis of his notebooks. In his own books, *Dialogue Concerning the Two Chief World Systems* (1632) and

Two New Sciences (1638), Galileo presents a series of arguments about the behavior of falling bodies. Strangely to our eyes, each consists of an extended conversation taking place over the course of several days among three men: Salviati, a stand-in for Galileo; Simplicio, who expresses the Aristotelian position and probably Galileo's earlier position (and, as his name implies, something of a simpleton); and Sagredo, an educated man of common sense. This literary format afforded Galileo much freedom to discuss politically and theologically sensitive issues, especially the Copernican system, without appearing to commit himself. If Salviati made an "impious" argument, Galileo could point out that he was only a fictional character whose views are not necessarily endorsed by the author. The format also allowed him to explore different ways of presenting his own arguments. Salviati's arguments thus do not necessarily capture Galileo's own actual thought process, but rather recapitulate his conclusions.

In both books, Salviati and Sagredo discuss several experiments that they claim to have conducted with bodies of different weights and compositions. During the discussion on the First Day in *Two New Sciences,* Salviati rejects Aristotle's apparent claim to have tested whether heavy objects fall faster than light ones. Sagredo then says:

> But I . . . who have made the test, assure you that a cannonball that weighs one hundred pounds (or two hundred, or even more) does not anticipate by even one span the arrival on the ground of a musket ball of no more than half [an ounce], both coming from a height of two hundred braccia [a braccio is about two feet] . . . the larger anticipates the smaller by two inches; that is, when the larger one strikes the ground, the other is two inches behind it.

Salviati adds that: "[I]t seems to me that we may believe, by a highly probable guess, that in the void [vacuum] all speeds would be entirely equal." Later, on the Fourth Day, he remarks:

> [E]xperience shows us that two balls of equal size, one of which weighs ten or twelve times as much as the other (for example, one of lead and the other of oak), both descending from a height of 150 or 200 braccia, arrive at the earth with very little difference in speed. This assures us that the [role of] the air in impeding and retarding both is small.[8]

Salviati may have been fictional, but he was clearly reporting on Galileo's own work. His claim to have performed an experiment, most historians believe, shows that Galileo did indeed drop objects of different weights to investigate and challenge Aristotle's account of motion. He apparently did so from towers—perhaps even the Leaning Tower—and to the discomfiture of his Aristotelian colleagues, who recognized from Galileo's other arguments that this spelled trouble not just for Aristotle's account of worldly motion, but for the rest of his system as well. It is true that some predecessors had also demonstrated shortcomings in Aristotle's account of motion, but Galileo did far more than they in showing how crucial a part of Aristotle's system this was, in developing an alternate account of motion, in developing the abstract thinking involved in this alternate one, and in illustrating its importance. Whether or not Galileo actually dropped balls from the Leaning Tower, he was the principal figure in developing an alternative to the Aristotelian theory of how bodies fall.

Viviani did well by his master. "*Se non è vero,*" as the Italians say, "*è ben trovato*" ("If it's not true, it might as well be") and we can be justified in calling it Galileo's Leaning Tower experiment.

But how and why did this experiment become so firmly entrenched in folklore as a turning point in the transition to modern science?

One reason is the strength of Viviani's account, which though brief is a captivating scene. While Viviani was generally careful and accurate, he was also writing for a particular audience—literary academics, clergymen, politicians, and other prominent nonscientists—who would not have cared for mathematics and technical details, and who would be engaged by a compelling story. "Viviani, one may assume," wrote science historian Michael Segre, "never imagined that some of his later readers would be incredulous historians of science."[9]

A second reason is a tendency for popular and even historical literature to pluck out a single episode to summarize and stand for a complex series of important events. In the case of the move from the Aristotelian to the modern framework, the Leaning Tower fits this bill admirably, though it does have the unfortunate effect of obliterating the context and implying that this experiment was the origin of Galileo's understandings of motion, and that considerations of motion were foremost in the clash between the two frameworks.

A final reason is our love of David and Goliath stories (at least, where David is one of our own) in which some reigning authority is exposed as illegitimate, humbled, and banished by a clever trick. Such stories seem to elevate our own wisdom.

EXPERIMENTS, LIKE OTHER TYPES of performances, have a creation or birth history that culminates in the first performance, and a maturation history that begins only then and covers everything that happens after—a biography, if you will. Like Eratosthenes' measurement of the earth's circumference, Galileo's experimentation on the motion of freely falling bodies was both something that he did at a certain time and place and a template for something that could be redone in different ways with different objects, technologies, and degrees of precision. Over time, Galileo's experimentation with falling bodies spawned a genre of experiments and demonstrations—offspring of the Leaning Tower.

The invention a dozen years after Galileo's death, for instance, of the air pump—which removes air from a chamber, making it possible to create an (imperfect) vacuum—allowed scientists, including Robert Boyle of England and Willem 'sGravesande of the Netherlands, to test Galileo's claim that bodies of unequal weight fall simultaneously in a vacuum.

Less scientifically exacting demonstrations of bodies falling in a vacuum continued to be popular even in the eighteenth century, when the new physics pioneered by Galileo had replaced that of Aristotle. Britain's King George III, for instance, insisted that his instrument makers stage a command demonstration featuring a feather and a one-guinea coin falling together inside an evacuated tube. One observer wrote:

> Mr Miller . . . used to tell that he was desired to explain the airpump experiment of the guinea and feather to Geo. III. In performing the experiment the young optician provided the feather, the King supplied the guinea and at the conclusion the King complimented the young man on his skill as an experimenter but frugally returned the guinea to his waistcoat pocket.[10]

Even in the twentieth century, some scientists still experimented with freely descending bodies, measuring their exact times of fall to test the equations for bodies accelerating in a resisting medium. One such experiment took place as recently as the 1960s, at the meteorological tower at Brookhaven National Laboratory on Long Island, by theoretical physicist Gerald Feinberg. "The main reason for bringing up a question which has so long been settled," wrote Feinberg, "is that the results of the theory are rather contrary to intuition, at least to the intuition of one brought up on Galileo's law." Equations in use for hundreds of years still needed corrections.[11] The Leaning Tower experiment, evidently, can still surprise us.

THE LEANING TOWER experiment addresses something fundamental: how objects—from cannonballs to feathers—behave under the influence of a force that affects us all. Its design is breathtakingly simple, with no mysterious fudge factors—one need not even have a watch. And it is definitive, leaving us with a particular kind of pleasure, which one might call "expected surprise." While we understand the truth of the Galilean framework, the Aristotelian framework is the one in which we live. If we lived on the moon, where there is no air resistance, the behavior of bodies falling in a vacuum would be familiar and the experiment would exert no disclosive power. But our daily experience leads us to expect objects to behave in the Aristotelian manner, and rewards us when we plan accordingly. When we pick up heavy objects, they tug down our hands more than lighter ones do, making us feel as though they ought to fall faster, as though they want to return to where they belong. For this reason, we still can delight in the perceptual display of that framework being violated, an experience that reinforces what we know intellectually. The pleasure involved recalls the game of *fort-da* described by Freud, in which the child made a small object disappear and then brought it back into view again; something brought the child endless delight in the object's return even though the child "knew" it was there all along.

Until recently, a few mysteries surrounded Galileo's experiments with falling bodies. One concerns his observation, in *On Mo-*

tion, that a less dense body when dropped first moves ahead of the more dense body, which eventually catches up. In the 1980s, science historian Thomas Settle had Galileo's experiments replicated with the aid of an experimental psychologist, and to his astonishment noticed the same thing. Further research convinced Settle that the heavier object makes the hand holding it more fatigued, which causes the experimenter to release it more slowly, even when he or she thinks they are dropping the objects simultaneously.[12]

Another recently clarified mystery involves the validity of Viviani's account, and why, if the experiment ever did occur at the Leaning Tower, Galileo never mentioned it in his own writings. In the 1970s, Galileo scholar Stillman Drake carefully examined Galileo's correspondence in 1641–42. Galileo, blind and under house arrest, was having Viviani read his correspondence and compose replies. Early in 1641, Galileo received several letters from his old friend and collaborator Vincenzio Renieri, who had just become a professor of mathematics at the University of Pisa, occupying the chair once held by Galileo himself. In one letter, Renieri wrote of carrying out an experiment in which he dropped two balls, one of wood and one of lead, from the "top of the campanile of the cathedral"—that is, the famous Leaning Tower. Galileo's reply is missing, but from Renieri's next letter it is clear that Galileo referred Renieri to his own account of experiments with falling bodies in *Two New Sciences,* and asked Renieri to repeat the experiment with bodies of different weights but the same material, to see if the choice of material would affect the results (it did not). Renieri's letter, moreover, seems to have jogged Galileo's memory about his own experiments in Pisa, which involved falling bodies of the same material, that he may have described to Renieri or at least to Viviani. If so, it would explain why Viviani may have been privy to a story that Galileo had long forgotten, and why Viviani, in his account, is quite specific that Galileo used balls of the same material. Viviani's stories have errors, to be sure, but they are usually minor errors of chronology, emphasis, or condensation. And why should Galileo have mentioned the Leaning Tower in his writings? He did mention "high places," and the fact that one of those high places might have been the Leaning Tower would have been only an accidental aspect of the experiment, having no bearing on the validity of the results. After mulling it over, Drake concluded that, in his letter to Renieri, Galileo probably described an experiment on falling bodies that he had performed at the Leaning Tower—which would have been the source of Viviani's story.[13]

Noted historian of science I. Bernard Cohen grew tired of replying "I don't know" to the questions, "Has anyone ever dropped two balls of different weights from the Leaning Tower?" and "What would happen if someone did?" At a meeting of the International Congress of the History of the Sciences in 1956, held at various locations in Italy, including Pisa, he paid a visit to the Leaning Tower, asked some colleagues and graduate students to shoo away passersby from a spot at its base, and climbed up its slippery, well-worn marble stairs and slanted floors. When he reached the top, he stretched out his arms, with some difficulty, over the lip at the southern edge and dropped two balls of different weights. They struck the ground at almost the same time—*boom, boom!*—before rapt onlookers. Rapt, surely, not because they were seeing something unexpected, but at least in part because they knew they were seeing something of historical significance: Galileo's famous Leaning Tower experiment, enacted for what might have been the very first time.

Interlude

EXPERIMENTS AND
DEMONSTRATIONS

Michelle Recreates
Galileo's Tower of Pisa Experiment

THAT IS THE TITLE of a life-size sculpture at the Boston Museum of Science. Michelle is an African-American preteen clad in overalls. She has stacked two chests on top of her bureau, clambered to the top, and now holds out a bright red softball in her left hand and a bright yellow golf ball in her right. She's poised to drop them when her mother walks in and stares up disapprovingly. Michelle's mother is thinking, according to the cardboard balloon above her head, "What in the world?!" Michelle herself is thinking, "I wonder which will hit the floor first?"

A caption says:

> How do falling bodies move? Will the softball hit the ground before the golf ball? Michelle, like Galileo 400 years before her, is seeing for herself. . . . "I'll see for myself." It's what you say when you don't want to just take someone's word for it.

This sculpture indicates the conceptual simplicity of Galileo's Leaning Tower experiment, expresses how legendary it has become, exhibits some of the simplifications of the legend, and illustrates some of the differences between experiments and demonstrations.

Michelle is doing an experiment, which is a kind of performance that discloses something for the first time. We stage performances when some issue has become important for us to clarify that cannot be clarified by reading any more on the subject, and to further our inquiry we have to plan, execute, observe, and interpret an action. In an experiment, we do not know how things will turn out in the end. This uncertainty makes us attend to the performance very carefully. And when the experiment shows what it does, it is not like learning the answer to a multiple-choice question, for we are transformed even when we are unsure of our next step. This is one of the differences Hardy pointed to between chess and mathematics; playing a game of chess does not change the rules, but a mathematical proof—or scientific experiment—changes science, for through it something new enters. Precisely because it does, our inquiry is not ended but modified and deepened.

Somehow, Michelle has become interested in falling bodies. Why, we do not know and is hard to imagine; probably something she read about Galileo puzzled her. She is serious enough to stage a little action, involving well-known ingredients, in the service of that inquiry. We also sense that whatever she finds will not be the end of it, and she will have more questions.

An experiment recapitulated becomes a demonstration. While an experiment is a performance whose authors and audience are the same—it is designed to disclose something to those who set it up and their community—a demonstration is a standardized performance whose authors and audience are different. If Galileo actually dropped balls of different weights off the top of the Leaning

Tower, it would have been as a demonstration—less for the purpose of disclosing something to himself than for convincing others. Today's landmark experiments turn into tomorrow's demonstrations. A demonstration is a recapitulation for a purpose, and depending on the purpose (inspiring a class, convincing colleagues, impressing reporters) one sets up the demonstration differently. The line between experiments and demonstrations is not always sharp, for in preparing and calibrating a new experiment one often comes to know what is to be revealed before the experiment "officially" begins—and the smart experimenter capitalizes on this knowledge to improve the experiment. And even demonstrations do not always go as planned, for instance, when disrupted by well-understood, mundane forces, or when something new inadvertently enters that one does not understand.

When we go to a science museum, we encounter demonstrations. The Boston Science Museum's falling-body demonstration, the "drop stop," has two tall Plexiglas cylinders next to each other, inside which are mechanical claws that can pick up objects of different weights inserted at the bottom, carry them to the top, and release them together. The paths of these objects are tracked electronically on the way down. Children frequently scour that floor of the museum for objects to insert, which delights the museum's curators but annoys the maintenance workers. The San Francisco Exploratorium has a different demonstration, consisting of a four-foot, freestanding Plexiglas cylinder mounted on an axle so that it can be rotated upside down. Two objects are inside the cylinder—a feather and a toy object of some sort, such as a rubber chicken—and it is attached to a vacuum pump that the museum-goer can turn on and off. The two objects are scooped up by a small shelf when the cylinder is rotated, and fall off the shelf when the cylinder is fully upside down. When you let air in the cylinder, the feather lags behind and takes several seconds to drift downward; but when the cylinder is evacuated, the two fall at about the same rate. The demonstration is so popular that the museum-goer invariably has to elbow a crowd of children out of the way to play with it.

Demonstrations tend to disguise the difficulty involved in conceiving, carrying out, and understanding experiments, creating a distance between audience and phenomenon not present in experiments. Demonstrations can also vastly simplify the experimental process through the use of modern equipment constructed with the "right" answer in view, even when rigged with imperfect results to promote verisimilitude.

Demonstrations, textbook accounts, and simulations can even misrepresent science by encouraging the sense that a scientific experiment is just an illustration of an already-formulated lesson— turning the experiment into a paint-by-numbers masterpiece, as it were—rather than a *process*. For this reason they can even diminish the beauty of science. Even though a scientific experiment may point to a simple fact, wrote science historian Frederic Holmes, it has been extracted from a "matrix of complexity" and inevitably introduces new dimensions of complexity.[1] So it was even with the Leaning Tower experiment: It took scientists a long time to figure out the importance of falling-body experiments, and they did not make science simpler but more complicated.

The Apollo 15 lunar feather-drop was, of course, a demonstration. As an experiment (except for the most rudimentary of explorations), it would have been inexcusably sloppy. Nobody measured the height from which the objects were dropped. Nobody cared that Scott was leaning over with his arms not parallel to the ground. Nothing was done to ensure that he released the objects at the same time. No measurement was made of the time of fall. As Commander Scott implied ("I guess one of the reasons we got here today was because of a gentleman named Galileo . . ."), scientists already knew the strength of the lunar gravity and the behavior of objects undergoing acceleration. If they had the slightest doubt about either, it would have been unwise to send a manned spacecraft to the moon in the first place.

Even as a demonstration, the Apollo 15 lunar feather-drop nearly met with disaster. In a trial run just a few moments before, Commander Scott discovered to his horror that static cling was making the feather adhere to his glove—but the demonstration worked miraculously when the camera was turned on. Fortunately, for thanks to the exotic location, the TV coverage, and the video clip NASA put on its Web site, the Apollo 15 experiment has gone on to become what is surely the most-watched scientific demonstration in history.

The Belled Plane. A late-eighteenth-century lecture-demonstration inclined plane from the Museum of the History of Science in Florence, Italy. A pendulum at the back was designed to ring a small bell at the end of each swing, marking equal intervals of time, and a set of moveable bells could be arranged along the plane so that a ball rolling past each of them would also create a ring. The demonstrator, by trial and error, could place the moveable bells so that a ball rolling down the plane would ring them in synchrony with the sounds made by the pendulum. Upon measuring the distances from the beginning of the roll to and between the positions of the bells on the plane, the demonstrator (and the audience) would find that the distances of the roll in successive equal time intervals were as the odd numbers beginning with one; that is, that the total distances were proportional to the squares of the total times. While this illustrates Galileo's Law, there is no evidence that Galileo ever constructed this particular version of the inclined-plane experiment.

Three

THE ALPHA EXPERIMENT:

Galileo and the Inclined Plane

—

SCIENCE TEACHERS CALL IT THE "ALPHA"—OR PRIMAL—
experiment. It is often the first experiment that students learn in
physics class. In many respects it was the first modern scientific experiment, in which an investigator systematically planned, staged,
and observed a series of actions in order to discover a mathematical law.

This experiment, which Galileo successfully carried out in
1604, introduced the concept of acceleration: the rate of change of
speed with respect to time. If the Leaning Tower experiment was a
demonstration that emerged from Galileo's studies of free fall indicating that bodies of unequal weight meeting negligible resistance
fall together, the inclined-plane experiment was a demonstration
that emerged from Galileo's studies of free fall illustrating the
mathematical law involved. Some mystery has surrounded this experiment, too, as the claims that Galileo made for it seemed far too
precise given the equipment available to him. But just as in the case
of the Leaning Tower experiment, recent historical work has un-

4 4 - **The Prism and the Pendulum**

covered surprises that have transformed our image of Galileo as an experimenter.

WHAT HAPPENS WHEN an object is released to fall freely? Does it pick up speed in a smooth fashion? Does it jump immediately to a "natural" uniform speed? Does it make some transition to a uniform speed? If these questions interest us, we might think to look and see what happens when, for instance, we release a coin or ball from our hands. But such bodies fall too quickly to track well. How can we arrange things to let us see with more precision?

Aristotle, as noted in the previous chapter, had examined bodies in motion, and had concluded, evidently from the way objects fall in water, that the speed of a falling body is uniform, proportional to its weight, and would be infinite in the absence of a resisting medium.

Galileo, however, became convinced that studying the motion of bodies in liquids obscured rather than clarified the issue. Like Aristotle, he found it too difficult to measure the paths of falling bodies directly because the eye is not swift enough, and existing timepieces were not accurate enough over short intervals. Instead of slowing down falling bodies by thickening the medium through which they passed, Galileo sought to dilute, as it were, the influence of gravity on their motion by rolling balls down inclined planes. This, he reasoned, might create a way to approximate the free fall of bodies. If the inclined plane was shallow, a ball would travel slowly; steepen it, and the ball travels faster. The steeper the incline, the more the ball's path would approach free fall. By measuring the rate at which objects rolled down an inclined plane, and how this rate changed as the incline steepened, Galileo hoped to solve the case of freely falling bodies (Figure 3.1).

By 1602, Galileo had built inclined planes into which he cut

*Figure 3.1.   The steeper
a plane's incline, the more
the motion of a ball
rolling down it approaches
free fall.*

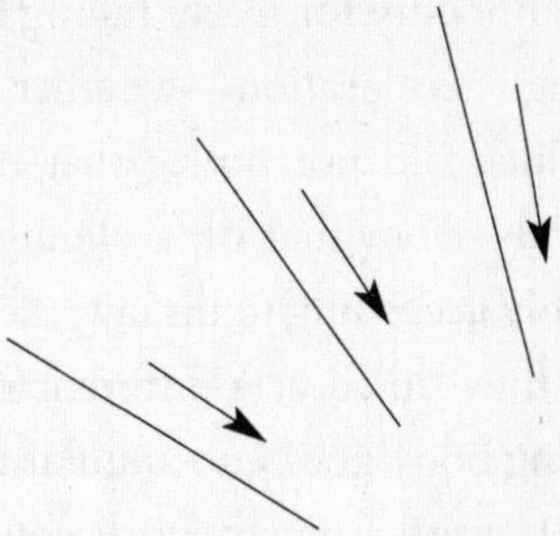

straight grooved tracks, and tried to measure how quickly balls rolled down them. But he was unsuccessful at obtaining useful results. He tried experimenting with pendulums, and learned much from them, for motion in the arc of a vertical circle is related to motion on an inclined plane. But he was unable to obtain the results he wanted. He began to appreciate the role of acceleration—that a falling body starts slowly and gains speed—and grew steadily more determined to find a mathematical description of it.

Galileo's notebooks and correspondence show that, by 1604, he had finally discovered the law of acceleration he was looking for as a result of his investigations of motion on inclined planes: The distance traversed by an object depends on the square of the time it is accelerated. If the time increases in even units (1, 2, 3, . . .), this means that the distance traversed by the object between each succeeding beat increases according to the odd-numbered progression (1, 3, 5, . . .). This is now known as Galileo's Law, $S \propto T^2$: The distance that a uniformly accelerated body moves from its starting point is proportional to the square of the time interval from its release (the modern equation is $d = \frac{1}{2}at^2$, or the distance covered by an object is equal to one-half the acceleration multiplied by the elapsed time squared). Galileo also discovered that this same law was true regardless of the angle of incline, and concluded that the same law that covered the acceleration of bodies sliding down inclined planes held not only for freely falling bodies, but also for any body undergoing acceleration—whether moving upward or downward. (Galileo did not notice that the motion of a rolling ball differs slightly from that of a sliding body. While both accelerate uniformly according to his law, the constant of acceleration is different and they do so at a different rate, for some of the energy of the rolling body goes into angular momentum.)

It was a momentous discovery. First of all, what Galileo did involved a change in what scientists looked at when they studied motion. Until then, one measured velocity in terms of space—how much ground was covered. Galileo was the first to realize that *time*, not space, was the independent variable to watch. We are so accustomed to doing this that it seems the natural way, but it is not, or at least it was not then. But more than that, Galileo showed that there existed no difference between the upward "violent motion" of heavy objects and their downward "natural motion." Both cases involved bodies being accelerated, hence the same mathematical law described their motion. In connection with Galileo's other work, this implied, again, that Aristotle's system could not easily be fixed, but had to be replaced.

Galileo reported his law in *Dialogue Concerning the Two Chief World Systems* (1632). Yet his brief account failed to convince some contemporaries, who complained of their inability to obtain the same results. In response to his critics, Galileo elaborated this in his next book, *Two New Sciences*. On the Third Day, upon hearing Salviati, the character who stands in for Galileo, mention Galileo's Law, Simplicio, the Aristotelian, objects:

> But I am still doubtful whether this is the acceleration employed by nature in the motion of her falling heavy bodies. Hence, for my understanding and for that of other people

like me, I think that it would be suitable at this place [for you] to adduce some experiment from those (of which you have said that there are many) that agree in various cases with the demonstrated conclusions.

Salviati finds this a reasonable request. After saying that, yes, he has indeed done the experiments and that they bear out the law in question, he describes to Simplicio the apparatus:

In a wooden beam or rafter about twelve braccia long, half a braccia wide, and three inches thick, a channel was rabbeted in along the narrowest dimension, a little over an inch wide and made very straight; so that this would be clean and smooth, there was glued within it a piece of vellum, as much smoothed and cleaned as possible. In this there was made to descend a very hard bronze ball, well rounded and polished, the beam having been tilted by elevating one end of it above the horizontal plane from one to two braccia, at will. As said, the ball was allowed to descend along the said groove, and we noted (in the manner I shall presently tell you) the time that it consumed in running all the way, repeating the same process many times, in order to be quite sure as to the amount of time, in which we never found a difference of even the tenth part of a pulse-beat.

Salviati tells Simplicio that, for instance, he rolled the ball a quarter of the length of the channel and found that it took exactly half the time; at other distances, the time elapsed according to the same ratio. "By experiments repeated a full hundred times," he says, "the spaces were always found to be to one another as the squares of the times. And this [held] for all inclinations of the plane, that is, of the channel in which the ball was made to descend." This is what we now call the law of uniformly accelerated motion.

Simplicio is persuaded: "It would have given me great satisfaction to have been present at these experiments. But being certain of your diligence in making them and your fidelity in relating them, I am content to assume them as most certain and true."[1]

GALILEO'S EXPERIMENT WAS different from Eratosthenes' measurement of the earth's circumference and his own Leaning Tower experiment. Those relied on equipment built for other purposes. The inclined-plane experiment, by contrast, required planning and building a distinctive piece of equipment with a specific function. Galileo's ingenuity lay not only in conducting the experiment itself, but also in designing the "stage" that made it possible. This stage creates a performance space in which a phenomenon—here, acceleration—can appear and be explored. Even a new and unexpected phenomenon: Galileo's *On Motion* manuscript shows that he began using inclined planes when he was still thinking that falling and even rolling bodies had a uniform motion. Once Galileo had designed his stage, it could be replicated by others to try their own performances, just as a playwright passes on a script for others to enact. For even though this experiment requires constructing its own stage, it still resembles Eratosthenes' measurement and the Leaning Tower experiment in that it can be reenacted in myriad ways.

Historians of science once believed that Galileo had discovered the law of uniformly accelerated motion, but were more skeptical than Simplicio about the fidelity with which he related the experiment. Their chief objection was to the way Galileo kept track of time. The device he used was a water timer, in which he measured the amount of liquid that flowed through a small pipe during the time of descent to deduce how much time had elapsed. These timers can be difficult to use accurately over short time intervals. Until recently, in fact, many science historians not only disbelieved but even ridiculed Galileo's claim to have used a water timer to measure "the tenth part of a pulse-beat," or approximately a tenth of a second. One of the more outspoken critics was Alexander Koyré of the École des Hautes Études in Paris, a Galileo specialist. Koyré had a Platonic view of science according to which it proceeded by theoretical reasoning and experiment was "theory incarnate." He allowed this prejudice to guide his reading of Galileo's writings, only took seriously Galileo's logical and mathematical arguments, and heaped scorn on his experimental work. In 1953, for instance, Koyré wrote of "the amazing and pitiful poverty of the experimental means at [Galileo's] disposal," and derided the inclined-plane experiment as follows:

> A bronze ball rolling in a "smooth and polished" wooden groove! A vessel of water with a small hole through which it runs out and which one collects in a small glass in order to weigh it afterwards and thus measure the times of descent . . . what an accumulation of sources of error and inexactitude! . . . It is obvious that the Galilean experiments are completely worthless: the very perfection of their results is a rigorous proof of their incorrection.[2]

Challenging Koyré, Thomas Settle, a poor, struggling graduate student in the history of science at Cornell, meticulously reconstructed the experiment in 1961 in the living room he shared with several other graduate students. Settle determined to use only "equipment and procedures which were available to Galileo or which were inherently no better than those he could muster." He chose a long pine plank, a set of wooden blocks, a flowerpot threaded by a small glass pipe plus a graduated cylinder (the water timer), and two different kinds of balls: a 2¼″ billiard ball, and a ⅞″ ball bearing. Making the experiment work right took a certain facility; Settle found that the operator "must spend time getting the feel of the equipment, the rhythm of the experiment. He must consciously train his reactions. And each day, or at the end of each break, he must be allowed a few practice runs to get warmed up." As Koyré had suggested, measuring the time of descent was indeed "the most difficult" aspect of the work. Nevertheless, in the end Settle learned that he could obtain excellent data in accord with Galileo's Law and concluded that Galileo's experiment "definitely was technically feasible for him," and even found that, with practice, his flowerpot timer could indeed be made precise to almost the tenth of a second claimed by Galileo. Settle published his reenactment in the journal *Science,* complete with diagrams and data tables (Figure 3.2). Despite some graduate-student bravado in his claim that replicating the experiment "essentially as Galileo described it" is "simple, straightforward, and easy to execute," this article remains an excellent guide to it.[3]

Settle's work established that the inclined-plane experiment could indeed demonstrate the law of uniformly accelerated motion. Nevertheless, many science historians still assumed that Galileo could not have experimented in the way he described; that is, he could not have used that method to establish his conclusions in the first place. These historians assumed Galileo had first discovered the mathematical law by some form of abstract reasoning, then built the device he described to illustrate it. The reason for their skepticism, again, was the water timers—they did not think he could have established the law with their aid.

In the 1970s, Galileo scholar Stillman Drake challenged this assumption as well. By carefully studying a page of Galileo's notebook, Drake concluded that Galileo actually had arrived at the law

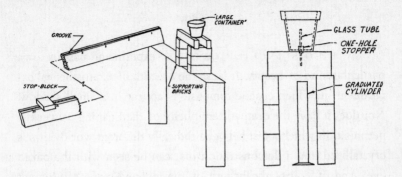

Figure 3.2. Diagram of Thomas Settle's apparatus re-creating Galileo's inclined-plane experiment.

using the inclined-plane method, but by marking out the time in a way that seems to have taken advantage of his strong musical training. A competent lute player, Galileo could keep a beat precisely; a good musician could easily tap out a rhythm more accurately than any water timer could measure. Drake determined that Galileo had set frets into the track of the inclined plane—movable gut strings of the kind used on early string instruments. When a ball was rolled down the track and passed over a fret, he would hear a slight clicking noise. Galileo, in Drake's speculative reconstruction, then adjusted the frets so that a ball released at the top struck the frets in a regular tempo—which for the typical song of the day was just over half a second per beat. Once Galileo had marked out fairly exact time intervals, thanks to his musical ear, all he would have to do would be to measure the distances between frets. These grew steadily longer as the ball picked up speed, illustrating the 1, 3, 5, . . . progression, and allowing him to compose the more elaborate experiment described in *Two New Sciences* and reconstructed by Settle.[4]

Galileo, in short, was a more skilled, and more ingenious, experimenter than science historians had given him credit for.

GALILEO'S INCLINED-PLANE experiment has its own particular kind of beauty. It lacks the breadth of Eratosthenes' experiment, in which cosmic dimensions appear in a small shadow. Nor does it have the dramatic simplicity of the Leaning Tower experiment, in which a test between radically different worldviews is crystallized into a demonstration that can be seen with the naked eye. And of course the beauty of the inclined-plane experiment does not lie in the mathematical law of accelerated motion that we find thanks to it, any more than the beauty of a Monet or Cézanne lies in the haystack or mountain painted. Rather, Galileo's inclined-plane experiment has a beauty of "pattern-emergence." Its beauty lies in the dramatic way in which a relatively simple apparatus allows a fundamental principle of nature to appear in what looks at first to be just a set of random and chaotic happenings—balls rolling down ramps. This was how the law first appeared to Galileo, and the way it is demonstrated to students today.

As one respondent to my poll wrote of his experience in reenacting Galileo's experiments, "The beautiful thing wasn't learning that gravity is 9.8 m/s^2, but in showing us that from a fairly simple setup we could quantitatively measure something important in physics."

THE NEWTON-BEETHOVEN COMPARISON

ONCE AFTER PLAYING BEETHOVEN'S last piano sonata, Opus 111, for some friends at a party, Werner Heisenberg told his spellbound audience, "If I had never lived, someone else would probably have formulated the principle of determinacy. If Beethoven had never lived, no one would have written Opus 111."[1]

And science historian I. Bernard Cohen cites a remark attributed to Einstein: "Even had Newton or Leibnitz never lived, the world would have had the calculus, but that if Beethoven had not lived, we would never have had the C-minor Symphony [the Fifth]."[2]

The Newton-Beethoven comparison, as it is often called, draws an elegant relation between the sciences and the arts with deep implications for the possibility of beauty in science. The usual argument contrasts the two, claiming that the products of science are inevitable but that those of the arts are not. The underlying assumption is that the structure of the world investigated by science is prefigured, and scientists work toward uncovering that already-existing structure. Sociologists of science call this the "paint-by-numbers" approach. Imagination, creativity, government interests, and social factors may affect the timing of when science gets done—how quickly and in what order the colors get filled in—but they cannot affect the structure of the final picture. Artists, on the other hand, are fully responsible for the overall structure of their works.

Philosopher Immanuel Kant also contrasted scientists and artists, but along different lines. According to Kant, "genius," despite the common romanticization of scientists like Newton, is not found among scientists, who are able to explain to themselves and others why they do what they do, but only among artists. While scientists can teach their work to others, artists produce *original* works, the secret of whose creation is unknown and unknowable. "Newton could show how he took every one of the steps he had to take in order to get from the first elements of geometry to his great and profound discoveries," Kant wrote, "not only to himself but to everyone else as well, in an intuitive[ly clear] way, allowing others to follow." Not so with Homer and other great poets. "One cannot learn to write inspired poetry, however elaborate all the precepts of this art may be, and however superb its models."[3]

Opposing the usual contrast, scientist Owen Gingerich has made a fascinating case in favor of an analogy, using a case study to show that scientists are partially responsible for the structure of their theories, meaning that the overall picture is not fully predetermined by nature. The Newtonian world system is not inevitable, he argued, for alternative explanations for celestial phenomena, in the form of Kepler's laws, can be derived from other sources, such as conservation laws. Gingerich's assertion of an alternative highlights the role of imagination and creativity in, and thus the singularity of, Newton's achievement. "Newton's *Principia* is a personal achievement that places him in the same creative class as Beethoven or Shakespeare," he concluded.

But Gingerich cautioned against drawing the Newton-Beethoven analogy too closely. "The synthesis of knowledge achieved in a major scientific theory is not fully the same as the ordering of the components in an artistic composition," he said. The scientific theory has a referent in nature, and is subject to "experimentation, extension, falsification." Scientific achievements can be legitimately and even inevitably paraphrased (who but historians read the *Principia* anymore?) in ways that artistic works cannot. And the way science progresses is different from progress in the arts. Nevertheless, Gingerich concluded, careful analysis of the Newton-Beethoven comparison—of the analogy and the contrast—allows us "to gain a more sensitive view of the nature of scientific creativity." While Kant's argument, and the traditional contrast, would seem to deny the possibility of beauty in scientific theories, Gingerich's seems to restore a place for it.[4] And French philosopher and scientist Jean-Marc Lévy-Leblond imagined in a thought experiment what the theory of relativity might look like if Einstein had never lived. The result has quite different terms, symbols, and ideas from the one we have now.[5]

If the issue is not theory but experimentation, the Newton-Beethoven comparison takes on yet another dimension. Experimentation is often regarded by outsiders as an automatic process involving minimal creative input. In this view, experimentation resembles the game show *Concentration*, which aired on U.S. television between 1958 and 1973. Contestants had to discover and interpret what lay behind the hidden faces of a set of blocks mounted in the wall. As the game proceeded, the blocks were rotated one at a time to reveal portions of a picture represented by words and symbols, which the contestants vied to decipher. The blocks were rotated by offstage technicians—the "experimenters"—activating hidden machinery. The mechanical process was of no real interest to the contestants, who only paid attention to the data on the surface.

As any experimenter can tell you, this view is wrong. Nothing is automatic or inevitable about a well-designed experiment. To appreciate this, however, requires that we look at experiments as a process, as well as a result. But how did the experiment get there? To understand this requires a story, almost a biography. In it there is inception, gestation, growth, and—with luck—maturity, followed by offspring. This process surely can involve what Kant calls genius, for which no rule exists beforehand.

Kant was surely right that style and tradition work differently within the sciences and the arts. Prism experiments may look back to Newton historically, precision measurements to Cavendish, light interference experiments to Young, and particle-scattering experiments to Rutherford. And historians who study long series of experiments by the same experimenter—such as Faraday, Volta, Newton, and Franklin—can often identify distinctive patterns in the way these scientists explored a phenomenon and devised new experiments to understand it. Still, an experiment is not recognizably a "Newton" or "à la Newton" in the same way that a painting may be recognizably a "Caravaggio" or "à la Caravaggio." Experimental work involves a different kind of ingenuity, also dependent on imagination and creativity, that is not inevitable and that creates its own kind of tradition of exemplars by virtue of opening up new domains of research.

The scientific imagination, like the artistic imagination, is disciplined. It works within a set of existing resources, theories, products, budget, and personnel, fashioning these elements into a performance that allows something new to appear. Of course, a larger budget and improved materials would be better! But the experimental imagination looks on the pool of existing resources not as limiting but enabling: As Goethe said, "Only in limitation is mastery revealed." In this respect, the Newton-Beethoven analogy is more comparison than contrast—and defines an unequivocal place for beauty in science.

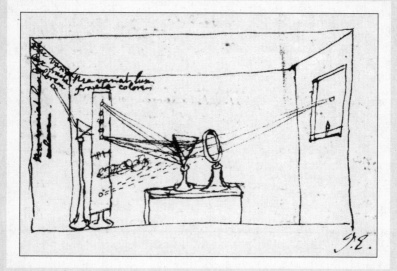

Isaac Newton's drawing of his experimentum crucis.

Four

EXPERIMENTUM CRUCIS:

Newton's Decomposition of Sunlight with Prisms

—

IN JANUARY 1672, ISAAC NEWTON (1642–1727) SENT A short note to Henry Oldenburg, the secretary of a newly established group of eminent scientists (or "philosophers," as they were then known) called the Royal Society of London. The Society had admitted Newton just one week before, its members impressed by his invention of an ingenious new type of reflecting telescope. Newton made Oldenburg a brazen claim. I have made a "philosophical discovery," Newton said, which was "in my judgment the oddest, if not the most considerable detection, which hath hitherto been made in the operations of nature."[1] Oldenburg might have been forgiven for thinking this preposterous, an arrogant claim by an overambitious youngster. And indeed Newton was a difficult person—combative, hypersensitive, and obsessively secretive. But this was no hyperbole.

A few weeks later, Newton sent the members of the Royal Society the description of an experiment that decisively showed, he said, that sunlight, or white light, was not pure as previously thought, but composed of a mixture of rays of different colors. Newton referred to this as his *experimentum crucis,* or "crucial experiment." His decomposition of light was at once a landmark in the history of science and a sensational demonstration of the experimental method. This experiment, one of Newton's many biographers wrote, "was as beautiful in its simplicity as it was effective in encapsulating Newton's theory."[2]

ISAAC NEWTON WAS BORN in Lincolnshire, England, in 1642, the same year Galileo died. He came into the world, perhaps appositely, on Christmas Day. From 1661 until 1665, Newton studied at Trinity College at Cambridge University. His was, another biographer asserted, "the most remarkable undergraduate career in the history of university education,"[3] for Newton discovered and thoroughly mastered on his own, and in the seclusion of his private notebooks, the new philosophy, physics, and mathematics that was being slowly and arduously forged by the most advanced scientists in Europe. In 1665, when Newton had graduated but was staying on for further studies, the Great Plague hit England and Cambridge University shut its doors for two years, sending Newton back to Lincolnshire. This enforced idle time among the fields and orchards of his mother's estate was not a setback to his education, but an unexpected blessing. It allowed Newton, then in his scientific prime, to think without interruption on numerous scientific topics on which he was already working at the cutting edge. Historians call this period of Newton's life his *annus mirabilis,* or "year of miracles," for in it he laid the foundations of many of his seminal ideas: in physics, the idea of universal gravitation (the inspirational falling-apple story, which comes to us via Newton's half-niece and Voltaire, supposedly took place at this time); in astronomy, the laws of planetary motion; in mathematics, calculus. During this time, Newton also began work on a revolutionary set of experiments in optics.

Optics, the study of light, was then of growing scientific importance. Since ancient times, thinkers had developed a basic knowledge of how light reflects and refracts (bends when passing through a transparent material). But prior to the seventeenth century, mirrors and lenses were of poor quality. Furthermore, their study was hampered by the prejudice that the images they produced were unworthy of serious examination because they were unnatural—how important could distorted and deceptive images be? But the invention of the telescope and microscope fueled the demand for better mirrors and lenses, which in turn increased interest in their manufacture and study. The new science, too, fostered the notion that optical distortions and transformations were not unnatural (like Aristotelian "violent" motions) but (again, like Galilean motion) just another arena governed by mechanical principles and mathematical laws that could be discovered through experimentation. Yet Descartes and other pioneers of optics in the seventeenth century still hewed to a view, found as far back as Aristotle, that white light is pure and homogeneous, with colors being a modification or "staining" of white light.

Cooped up on his mother's estate, while disease raged in the cities, Newton transformed one of her rooms into an optics laboratory, sealing it from light except for a tiny hole to the outside. There he spent day after day absorbed in experiments. Wrote one associate, "[T]o quicken his faculties and fix his attention [he] confined himself to a small quantity of bread, during all the time, with a little sack and water, of which, without any regulation, he took as he found a craving or failure of spirits." Newton's principal tool was a prism, a popular curiosity of the time highly regarded for its ability to transform white light into various colors. But Newton transformed the toy into a powerful instrument for scientific investigation in his study of light.

A common stereotype, inflicted on generations of schoolchildren, is that the scientific method is a robotic enterprise: forming, testing, and reforming hypotheses. A vaguer-sounding but more accurate account of what scientists do would be to say they "look at" a phenomenon—they examine it from different angles, understanding it by tweaking it this way and that to see what happens. In his converted laboratory, Newton "looked at" light, using various configurations of prisms and lenses, and eventually arrived at the conclusion that white light was not pure but a mixture of light of different colors. Newton would write later, "For the best and safest method of philosophising seems to be, first to inquire diligently into the properties of things, and establishing those properties by experiments and then to proceed more slowly to hypotheses for explanations of them."[4]

But for several years Newton told few people about this work. When he returned to Trinity College when it reopened in 1667, he attended lectures on optics delivered by Isaac Barrow, the first occupant of the Lucasian chair of mathematics at Cambridge (a famous chair, whose later occupants included Paul Dirac and Stephen Hawking), proofread Barrow's lecture notes, and in 1670 succeeded Barrow as Lucasian Professor. The post required him to wear a scarlet robe to indicate his elevated status over other faculty. It also required him to deliver a lecture to students at least once a week, in Latin, on some topic pertaining to mathematics. Newton chose optics, which allowed him to mix mathematics and experimental science and to "bring the principles of this science to a more strict examination." These lectures were sparsely attended. An associate remarked that "So few went to hear him, & fewer that understood him, that oftimes he did in a manner, for want of hearers, 'read to the walls.' "[5] Literally read to the walls—not a single person attended his second lecture.

In 1671, Newton presented members of the Royal Society with a telescope he had invented based on his optical studies. The Royal Society had been established just over ten years earlier, as the Royal Society of London for Improving Natural Knowledge; its motto, inscribed on its coat of arms, was the Latin expression "*Nullius in verba*," which is traditionally translated as "Don't take anyone's word for it." The Royal Society met weekly, discussing and analyzing members' papers. This format was crucial to the stimulation of research and the professionalization of science, for it streamlined the process by which scientific information was disseminated and defended; one could focus on researching a specific topic and report one's results in a letter. These letters were published in what was initially called the Society's *Correspondence* and later the *Philosophical Transactions*, a forerunner of the modern scientific journal. When Newton joined, few members had heard of him. Nevertheless, his telescope was a sensation. Only six inches long, it was ingeniously designed and carefully built, and the equal of many much larger telescopes. Several members began trying to construct one for themselves, and they soon invited Newton into their ranks.

Newton's first formal submission to the Society was the letter that fulfilled his bold promise to Oldenburg to relay news of the "oddest" philosophical discovery yet made about nature's operations. This paper is often cited as a masterpiece of scientific literature and a model of scientific writing. It provides an excellent description not only of the crucial experiment itself but also of the thought process that led up to it—and a sharp reader will notice, between the lines, the sheer joy that Newton took in his investigations. The paper begins as follows:[6]

To perform my late promise to you, I shall without further ceremony acquaint you, that in the year 1666 . . . I procured me a Triangular glass-Prisme, to try [to test] therewith the celebrated *Phenomena of Colors*. And in order thereto having darkened my chamber, and made a small hole in my window-shuts, to let in a convenient quantity of the Suns light, I placed my Prisme at its entrance, that it might be thereby refracted to the opposite wall. It was at first a very pleasing divertisement, to view the vivid and intense colors produced thereby.

Others might have fallen into the temptation to pay attention only to the beguiling rainbow-like play of colors. But Newton was looking at what was happening from as many angles as possible. He saw beyond the colors to the shape they were assuming. "I became surprised to see them in an *oblong* form; which, according to the received laws of Refraction, I expected should have been *circular*."

Why was Newton surprised? In the prevailing view of Descartes and others, prisms somehow modified or stained white light to produce the spectrum. If so, a pencil-thin beam should emerge from the prism with the same circular outline that it had when it entered. Instead, Newton saw, the image was shaped like a racetrack, with semicircular curves at the top and bottom connected by straight sections (Figure 4.1); the colors were arranged in horizontal bands, with blue at one end and red at the other. Newton also noticed a second puzzling feature: While the straight sections of the image were crisp, the curves at either end—blue or red—were blurred. This, plus the "extravagant" discrepancy between the length and the width—the former was about five times longer than the latter—"excited me to a more than ordinary curiosity of examining, from whence it might proceed."

Newton next describes his attempts to determine why the image

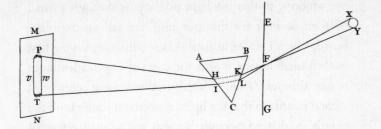

Figure 4.1.   Newton's diagram of the oblong shape produced by a beam of sunlight after passing through a prism.

had acquired such an unexpected outline simply after passing through the prism. Trying to see if he could affect the shape of this outline, he tried using prisms of different thicknesses, and tried passing light through different parts of the prism. He rocked the prism back and forth on its axis. He changed the size of the hole in the window, and tried putting the prism outside in the sunlight so that the beam came through the prism before passing through the hole in the window. He checked to see if imperfections in the glass of the prism might be responsible. None of this affected the shape of the outline. Its puzzling oblong shape remained, and each color was always refracted—sent off at an angle while passing through the prism—in the same way.

Newton remembered times when he had seen "a tennis-ball, struck with an oblique racket," following an arc through the air. Maybe, he began to suspect, the shape of the spot could be explained if the prism somehow caused the light rays to travel in curved paths in the vertical direction. Thus drove him to another set of experiments.

The gradual removal of these suspicions at length led me to the *Experimentum Crucis*, which was this: I took two boards, and placed one of them close behind the Prisme at the window, so that the light might pass through a small hole, made in it for that purpose, and fall on the other board, which I placed at about 12 foot distance, having first made a small hole in it also, for some of the Incident light to pass through. Then I placed another Prisme behind this second board, so that the light, trajected through both the boards, might pass through that also, and be again refracted before it arrived at the wall. This done, I took the first Prisme in my hand, and turned it to and fro slowly about in Axis, so much as to make the several parts of the Image, cast on the second board, successively pass through the hole in it, that I might observe in what places on the wall the second Prisme would refract them. And I saw by the variation of those places, that the light, tending to that end of the Image, towards which the refraction of this first Prisme was made, did in the second Prisme suffer a Refraction considerably greater than the light tending to the other end.

Newton's own diagram for his *experimentum crucis*, which he drew on a piece of paper in his first lectures on optics, is shown in Figure 4.2. A pencil-like beam of light coming in through a hole in the window passed through a first prism and fanned out against a board a dozen feet away. In fanning, it threw out a rainbow-like display of colors—oblong in the vertical dimension, but with horizontal bands of color from red to blue. Anyone who had played with prisms had seen this, though not necessarily realizing the significance of the shape. But what Newton did next was novel: He added a second prism and board. He drilled a hole in the board, passed part of the oblong band of light through it to another prism on the other side, and then directed that beam against *another* board. By swiveling the first prism, he could maneuver the oblong band up and down so that light of different colors would pass through the hole, through the second prism to the second board. He then looked carefully at what happened.

Newton noticed that blue light, which was greatly refracted by the first prism, was also greatly refracted by the second prism as well; similarly, red light, less refracted by the first prism, was less refracted by the second. He also noticed that *how* these were refracted did not depend on the angle of incidence (the angle at which they struck the surface of the prism). Newton concluded that the degree by which the rays were refracted—their "refrangibility," after the Latin word *refrangere*, "to break off"—was a property of the rays themselves and not of the prism. The rays kept their refrangibility while passing through the two prisms. The prisms did not modify the light rays but only sifted them according to their refrangibility.

Newton now held the answer to his initial question. The rainbow image was shaped like a racetrack because the prism spreads out the beam of light in a way dictated by the behavior of its individual component colors. If the axis of the prism is horizontal, the prism maintains the beam at the same width but spread it out vertically. The vertical tips of the oblong shape are blurry because there are fewer rays at the extreme top and bottom. Newton wrote: "And

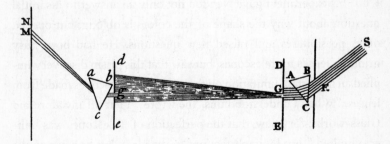

Figure 4.2.   Newton's diagram of his experimentum crucis,
from his lectures on optics.

so the true cause of the length of that Image [the oblong shape] was detected to be no other, than that *Light* consists of Rays differently refrangible, which, without any respect to a difference in their incidence, were, according to their degree of refrangibility, transmitted towards divers parts of the wall."

What was so crucial about this experiment, out of the hundreds that Newton had performed, many of which showed him similar effects? His own confidence in his conclusions about this experiment was based, not on it alone, but on all of his other attempts to look at light with prisms and lenses. But Newton saw no reason to insist that his colleagues follow his own extensive path of inquiry. All it should take to set his colleagues on the right track would be a single one. Thus there was a certain theatricality to the *experimentum crucis;* it was a demonstration or recapitulation of what by now he had already learned how to do. The purpose of the demonstration was to persuade colleagues, so it needed to be simple, with readily available instruments, and show the result cleanly and vividly to maximize the impact. As he would write later to someone struggling to recreate his experiments, "Instead of a multitude of things try only the Experimentum Crucis. For it is not [the] number of Expts, but [their] weight [is] to be regarded; & where one will do, what need of many?"[7]

This experiment gave Newton not only an answer to his initial question about why the shape of the colors is oblong, but opened other possibilities and raised new questions. He had been busy grinding lenses for telescopes, but saw that this prism discovery implied an important limitation on the quality of telescopes made from lenses. "When I understood this," he wrote, "I left off my aforesaid Glass-works; for I saw, that the perfection of Telescopes was hitherto limited," not by imperfections in the glass, but because of the fact "that Light it self is a *Heterogeneous mixture of differently refrangible Rays.*" Lenses focus by bending or refracting light; but because different kinds of light refract by different amounts, even a perfect lens could not collect all the rays at one point. A more effective way to focus light for a telescope, Newton realized, would be to use mirrors rather than lenses, since when mirrors bounce or reflect light to focus it, the angle at which different kinds of light reflect is always the same. Newton said that he promptly set out to build a telescope that used mirrors, but his telescope making was interrupted by the plague. In 1671, he finally built one he was confident in and proud of—proud enough that he was able to overcome his usual obsessive secretiveness and show it to the Royal Society.

Newton laid all this out in the first half of the paper. In the second half, he discussed several implications of his discovery. A first was that the refrangibility of light was not a property caused by the prism through some form of modification, as Descartes and most other authors on the subject believed: "Colours are not *Qualifications of Light*, derived from Refractions, or Reflections of natural Bodies (as 'tis generally believed,), but *Original* and *connate properties*, which in divers Rays are divers. . . ." A second implication was that "To the same degree of refrangibility ever belongs the same colour, and to the same colour belongs the same degree of refrangibility." A third was that the refrangibility or color of a ray is unaffected by the substance it passes through. Newton had looked at this point very carefully:

> The species of colour, and degree of Refrangibility proper to any particular sort of Rays, is not mutable by Refraction, nor by Reflection from natural bodies, nor by any other cause, that I could yet observe. When any one sort of Rays hath been well parted from those of other kinds, it hath afterwards obstinately retained its colour, notwithstanding my utmost endeavours to change it. I have refracted it with prisms, and reflected it with bodies, which in day-light were

of other colours; I have intercepted it with the coloured film of air, interceding two compressed plates of glass; transmitted it through coloured mediums, and through mediums irradiated with other sorts of rays, and diversly terminated it; and yet could never produce any new colour out of it.

Newton comes to the remarkable conclusion that white light is not original but compound, a fact that he had confirmed in some of his experiments by using additional prisms and lenses to recombine light that he had earlier separated:

> But the most surprising and wonderful composition was that of Whiteness. There is no one sort of Rays which alone can exhibit this. 'Tis ever compounded and to its composition are requisite all the aforesaid primary Colours, mixed in a due proportion. I have often with admiration beheld, that all the colours of the prism being made to converge, and thereby to be again mixed, as they were in the light before it was incodent upon the prism, reproduced light entirely and perfectly White. . . . Hence, therefore, it comes to pass, that Whiteness is the usual colour of light; for light is a confused aggregate of rays indued with all sorts of colours, as they were promiscuously darted from the various parts of luminous bodies.

Newton's "surprising and wonderful" discovery ignited new insights into what had been deep mysteries. In the rest of the article he addresses some of these one by one, solving with ease puzzles that had baffled his colleagues. How do prisms work, and how do they make the oblong shape of the spot they produce? They do not transform, but instead sift light, separating it into bands of like refrangibility. Imagine (this is not Newton's image) a pack of runners, each of which is able to turn a corner at a different angle. Though they keep together when moving in a straight line, at the first sharp corner they will fan out into a band. How do rainbows form? Newton explains this as raindrops acting like a cloud of tiny prisms, refracting the light of the sun behind them. What about those "odd phenomena" involving colored glass and other materials in which the same stuff gives off different colors? These are "no longer riddles," Newton says, for they are materials that reflect and transmit different kinds of light in different conditions.

Newton accounted for an "unexpected experiment" made by Robert Hooke, the Royal Society's Curator of Experiments. Hooke had shined light through a jar of red liquid and a jar of blue liquid. Each one let light through—but when he tried to shine light through both jars together, they blocked all light. Hooke had not been able to explain this: Why, if each individual jar allowed light to pass through, would the combination block all light? Hooke's puzzlement, Newton said, was evidently due to the assumption that light was a uniform substance; instead, light was composed of many types of rays. The blue jar let through one type but blocked all others; the red jar let through a second type, and blocked all others. Because the two jars did not allow the same type of light to pass, "no rays could pass through both."

Newton was now also able to explain the color of natural bodies—they reflect "one sort of light in greater plenty than another"—and he described his own experiments in a darkened room, in which he cast light of different colors on various objects, finding "by that means any body may be made to appear of any color." Are there colors in the dark, and is color a property of objects? No—color is a property of the light that shines on them.

Newton ended the letter with some suggestions for experiments his colleagues could make, though he warns that these experiments, like the *experimentum crucis*, are highly sensitive. The prism has to be of high quality or the light reaching the second prism will be impure, and the room must be absolutely dark lest light mix with the colors and confuse the issue. This latter feature makes the *experimentum crucis* more difficult to re-create than it may seem for high school science-education classes, however temptingly accessible and vividly instructive it might appear. Newton concluded:

> This I conceive is enough for introduction to experiments
> of this kind; which if any of the Royal Society shall be so
> curious as to prosecute, I shall be very glad to be informed
> with what success: that if any thing seem to be defective, or
> to thwart this relation, I may have an opportunity of giving
> further direction about it; or of acknowledging my errors,
> if I have committed any.

Newton's letter reached Oldenburg on February 8. As luck would have it, Oldenburg was preparing for a Royal Society meeting later that day and was able to put it on the agenda. Those present first sat through a letter on the possible influence of the moon on barometric readings, then another on the effects of a tarantula's sting, before hearing Newton's contribution. The Society was greatly impressed. Oldenburg reported that "[T]he reading of your discourse concerning Light and Colours was almost their only entertainment for that time. I can assure your, Sir, that it there mett both with a singular attention and an uncommon applause."[8] Oldenburg also mentioned that the members had directed him to publish it as soon as possible in the *Philosophical Transactions,* and it appeared in the next issue later that month.

NOT ONLY WAS NEWTON'S *experimentum crucis* a beautiful experiment, and his letter about it in the *Philosophical Transac-*

tions the very model of a scientific paper, but it also spawned what is surely the first "journal controversy," in which scientists argue heatedly back and forth about an issue. Newton's experiment, challenging as it did the orthodoxy of the time, according to which prisms created colors by modifying white light, created a stir in the Royal Society and among other scientists, especially in France.

Without trying to replicate the *experimentum crucis*, Robert Hooke had dismissed Newton's letter a week after reading it with some rash and incorrect criticisms about the hypotheses Newton seemed to be making. Newton rose to the occasion, brilliantly displaying his combative talents in the exchange of letters that followed, recapitulating and elaborating his arguments, which included one of the most sarcastic put-downs in history. This made use of the fact that Hooke was so short and hunched (partly exacerbated by the type of exacting bench work that he did) that he resembled a dwarf. In one letter that fairly dripped with false flattery, Newton praised Hooke's contributions to his own work with the words, "If I have seen further it is by standing on [the] shoulders of Giants."[9] This famous remark is now often cited as gallant and humble when in fact it nastily ridiculed Hooke.

Scientists in France took longer to convert. One was an elderly professor at the College of English Jesuits at Liège named Francis Hall, though he called himself Linus in correspondence. In the fall of 1674, Linus—who was pushing eighty—wrote to Oldenburg claiming that, in experiments with prisms that he had conducted thirty years earlier, he never observed an elongated outline on sunny days, claiming the elongation of the image that Newton saw was due to the effects of clouds. Newton, who considered Linus incompetent, did not deign to answer. Oldenburg, however, ordered Hooke to stage a demonstration of Newton's *experimentum crucis* at a Royal Society meeting in March 1675. The weather, alas, did not cooperate, and in view of Linus's remarks it was thought pointless to carry out the experiment on a cloudy day. Linus died in the fall of that year, but his cause was honored by a devout pupil who expressed confidence that his master would be vindicated the next time the Royal Society tried the experiment on a sunny day.

Hooke again made plans for a demonstration at the Royal Society, and what Newton referred to as "ye Experiment in controversy" was rescheduled for 27 April 1676 (a sunny day, as it turned out). Although Newton was not present—he generally shunned such public occasions—it was a landmark in the dawn of modern science, for it was the first experiment planned and executed by a scientific society to obtain the decisive answer to a pressing controversy. The official records of the Royal Society state:

> The experiment of Mr. NEWTON, which had been contested by Mr. Linus and his fellows at Liege, was tried before the Society, according to Mr. NEWTON's directions, and succeeded, as he all along had asserted it would do: and it was ordered, that Mr. OLDENBURG should signify this success to those of Liege, who had formerly certified, that if the experiment were made before the Society, and succeeded according to Mr. NEWTON's assertions, they would acquiesce.[10]

Some French critics held out a few years longer. A French Jesuit named Anthony Lucas tried the *experimentum crucis* but found red rays among the purple; another found red and yellow among the violet. Newton stopped responding, writing that "[t]his is to be decided not by discourse, but new Tryall of ye experiment."[11] He had already delivered warnings about what could go wrong with the experiment. Like any complexly performing device, an experiment can be arranged incorrectly—but when done right, it shows what went wrong in the incorrect trials; it provides its own criteria of success.

NEWTON'S *EXPERIMENTUM CRUCIS* provided the world with many things at once: a piece of information, a set of tools and techniques, even a moral lesson. It owes its beauty to each of these. Newton's experiment disclosed a piece of truth about the world with astounding simplicity and ingenuity: Who would have thought, after using a prism to split a beam of white light into a rainbow, to pick a portion of that and send it through *another* prism? With that configuration, no further manipulations were necessary to show Newton's colleagues that white light is composed of rays of different colors with different degrees of refraction.

The experiment allowed us to understand many puzzling phenomena of light, and gave us techniques for separating light of different colors and of building better telescopes. Newton's insight erupted like a firecracker, shooting connections in many different directions.

Finally, Newton's *experimentum crucis* was a moral lesson for scientists. It said, in effect: "This is the way to go about understanding a puzzling phenomenon. Experiment long and hard. Then pick out the most economical and vivid demonstration you can find, point out the ways it can go wrong, and show what new connections it makes possible."

Thus its beauty has nothing to do with the prettiness of the colors themselves; Newton, like Eratosthenes with his shadows, looked beyond the colors to what was making them behave the way that they do. But like Galileo's inclined-plane experiment, Newton's *experimentum crucis* revealed something about the nature of experimentation itself. What's distinctive about the *experimentum crucis* is that it has a kind of moral beauty.

By 1721, when a second French edition of Newton's *Opticks* was scheduled to be published in Paris (the first having appeared in 1704), its French publisher, Varignon, wrote to Newton, "I have read the Opticks with the greatest delight, and all the more so because your new system of colours is firmly established by the most beautiful experiments." Varignon asked Newton for a drawing to put at the top of the first page that might symbolize the contents.

Newton chose a drawing of the *experimentum crucis* with a laconic caption: "Light does not change color when it is refracted." It was an elegant symbol of what, in Newton's hands, became the science of optics itself.

DOES SCIENCE DESTROY BEAUTY?

When I heard the learn'd astronomer,
When the proofs, the figures, were ranged in columns
    before me,
When I was shown the charts and diagrams, to add,
    divide, and measure them,
When I sitting heard the astronomer where he lectured
    with much applause in the lecture-room,
How soon unaccountable I became tired and sick,
Till rising and gliding out I wander'd off by myself,
In the mystical moist night-air, and from time to time,
Look'd up in perfect silence at the stars.

—WALT WHITMAN

TO LOVERS OF BEAUTY, Newton brought not peace but a sword.
Early philosophers, poets, and artists viewed light as having a
special status among all the world's phenomena. Plato compared the
sun and its rays to the Good—the highest form—for it not only nur-
tured but illuminated everything. Followers in the Platonic tradition,
including St. Augustine, Dante, Grosseteste, and St. Bonaventure,
saw a special tie between light and beauty or being; light was the
principle of all visible and sensuous beauty, and beautiful in itself. It

illuminated the world that God had made; it was epiphany. Light, naturally, had a special status for painters, too; they treated light around Newton's time as "an act of love," in the words of Kenneth Clark, for light seemed to spread, brighten, and intensify the world.[1]

But the rise of modern science, and especially the work of Newton, posed a challenge to this view. Light suddenly lost its status as the principle of epiphany. It was no longer the world illuminating itself, through light, for the benefit of humans; it was rather the human mind reaching out to illuminate the world. Light became just another phenomenon governed by knowable mechanical and mathematical laws.[2] The measure of poetic response to the new science was what poets thought Newton had done to that treasure chest of colors, the rainbow.

To some poets and artists of the eighteenth and early nineteenth centuries, Newton was the enemy. He seemed to have transformed the rainbow, and other manifestations of color, into an exercise in mathematics. Keats was one. In 1817, Keats lamented that "the rainbow is robbed of its mystery"; and at a party, Keats and writer Charles Lamb chastised their host, British painter B. R. Haydon, for inserting Newton's head into one of his paintings, claiming that Newton "had destroyed all the poetry of the rainbow, by reducing it to the prismatic colors."[3] A year and a half later, still agitated, Keats addressed the subject again in his poem "Lamia" (1820) using the then-common practice of referring to science as "natural philosophy," and "awful" to mean "full of awe":

> Do not all charms fly
> At the mere touch of cold philosophy?
> There was an awful rainbow once in heaven:
> We know her woof, her texture; she is given
> In the dull catalogue of common things.

Philosophy will clip an Angel's wings,
Conquer all mysteries by rule and line,
Empty the haunted air, and gnoméd mine—
Unweave a rainbow . . .

That same year also saw the publication of Thomas Campbell's poem "To the Rainbow":

Can all that optics teach, unfold
    Thy form to please me so,
As when I dreamt of gems and gold
    Hid in thy radiant bow?

When Science from Creation's face
    Enchantment's veil withdraws,
What lovely visions yield their place
    To cold material laws!

Poet William Blake depicted a naked Newton in one drawing as a bearded man measuring things out precisely with a compass, and wrote:

The atoms of Democritus
And Newton's particles of Light,
Are sands upon the Red Sea shore:
Where Israel's tents do shine so bright.

In his works *Theory of Colors* and *Contributions to Optics*, Johann Wolfgang von Goethe even went so far as to try to develop an explicitly counter-Newtonian science of color based solely on how it is perceived. Goethe conducted a remarkable series of experiments himself, and managed to describe and explain aspects of color perception that Newton had not noticed. Goethe's work strongly influenced many artists, including the painter J.M.W. Turner.

But another set of artists approached Newton's achievement with different eyes. Newton himself did not seem to be much of an art connoisseur; he once dismissed sculptures as "stone dolls," and liked to cite Isaac Barrow's opinion of poetry as "a kind of ingenious nonsense." Still, many artists found him to have opened up new realms of beauty. One was the British poet James Thomson, who, as Marjorie Nicolson pointed out, along with several peers learned to see rainbows and sunsets with "Newtonian eyes":[4]

Even now the setting sun and shifting clouds,
Seen, Greenwich, from thy lovely heights, declare
How just, how beauteous the refractive law.

As M. H. Abrams wrote, Thomson seemed to believe that "Newton alone has looked on beauty bare."

The split between the Romantic poets of the eighteenth and early nineteenth centuries represents a split that is still with us, between those for whom inquiry and investigation destroy beauty, and those for whom they deepen beauty. Physicist Richard Feynman was once taken to task by an artist friend, who claimed that while artists can see the beauty of a flower, a scientist would take it apart and turn it into a cold, lifeless thing. Feynman, of course, knew better. He retorted that, as a scientist, he was able to see not less but more beauty in the flower. He could appreciate, for instance, the beautiful, complicated actions inside the flower's cells, in its ecology, and in its role in evolutionary processes. "Science knowledge," Feynman said, "only adds to the excitement and mystery and the awe of a flower."[5]

Learning about such things no more diminishes an appreciation of the flower than learning acoustics diminishes one's appreciation of Vivaldi's *Four Seasons*. Maintaining our sense of wonder at the world is not achieved by retreating from science but by engaging with it. The antidote to the learned astronomer is the good one— the one who continues to share this wonder.

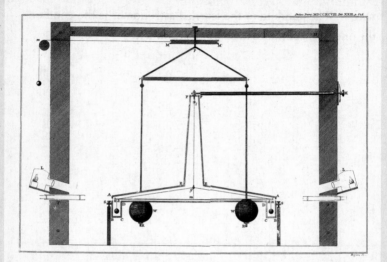

Henry Cavendish's equipment to measure the density of the earth.

Five

Cavendish's Austere Experiment

—

THE ENGLISH SCIENTIST HENRY CAVENDISH—ONE OF the greatest chemists and physicists of the eighteenth century—was also one of its strangest. Fortunately for him and for science, his aristocratic background and inherited wealth gave him the means to indulge his interests in his own way. As a result, he managed to carry out an extraordinary experiment whose precision could not be significantly bettered for a century.

Cavendish (1731–1810) had a nervous, squeaky voice, dressed in quaint clothes that were literally fifty years out of date, and shunned people as much as possible. His first biographer, a Royal Society scientist named George Wilson, remarked that colleagues said Cavendish costumed himself like their grandfathers, including a cocked three-cornered hat, and that he was "shy and bashful to a degree bordering on disease."[1] When he had to suffer being introduced to people, Cavendish often silently gazed over their heads, if he did not bolt the room in distress. Sometimes he would stand, frozen, at the threshold of a crowded room, literally unable to make an entrance. When riding in a carriage he huddled in a corner to prevent himself from being seen through the open windows. On his daily constitutional, he always took the same route at the same time, walking in the middle of the road to avoid chance encounters. When he realized that his neighbors had figured out his routine and were gathering to stare at the local eccentric, Cavendish changed his schedule to take his stroll under cover of night. The only portrait of Cavendish ever made had to be done secretly. His acquaintances, knowing he would be too shy to agree, secretly invited a painter to a Royal Society dinner and sat him near the end of the table so as to get a good look at Cavendish's face. *Cavendo tutus* ("Be safe by being cautious") was the motto of the Cavendish family, but Henry's behavior took this counsel to a pathological extreme.

Cavendish, whose mother died when he was two years old, was especially afraid of women. To avoid having to deal with his female housekeeper, he always left written instructions for the next day's work and meals on a table before retiring to bed. After he accidentally encountered the housekeeper on the stairs, he had a back stairway installed to prevent this from happening again. And in the recollection of a Royal Society acquaintance:

> One evening we observed a very pretty girl looking out from an upper window on the opposite side of the street, watching the philosophers at dinner. She attracted notice, and one by one we got up and mustered round the window to admire the fair one. Cavendish, who thought we were looking at the moon, bustled up to us in his odd way, and when he saw the real object of our study, turned away with intense disgust, and grunted out Pshaw![2]

Cavendish was utterly methodical in life and work. He always ate the same thing for dinner: a leg of mutton. His daily routines,

Wilson noted, were carried out according to a law that was "as inflexible and imperative as that which rules the motions of the stars":

> He wore the same dress from year to year, taking no heed of the change in fashions. He calculated the advent of his tailor to make a new suit of clothes, as he would have done that of a comet. . . . He hung up his hat invariably on the same peg, when he went to the meetings of the Royal Society Club. His walking-stick was always placed in one of his boots, and always in the same one. . . . Such was he in life, a wonderful piece of intellectual clockwork; and as he lived by rule, he died by it, predicting his death as if it had been the eclipse of some great luminary (which in truth it was), and counting the very moment when the shadow of the unseen world should enshroud him in its darkness.[3]

Wilson, a careful and perceptive writer, was strongly ambivalent about his biographical subject. When forced to assess Cavendish as a person, Wilson agonized and struggled heroically to produce the following remarkable evocation of this strange and brilliant man:

> Morally [his character] was a blank, and can be described only by a series of negations. He did not love; he did not hate; he did not hope; he did not fear; he did not worship as others do. He separated himself from his fellow men, and apparently from God. There was nothing earnest, enthusiastic, heroic, or chivalrous in his nature, and as little was there anything mean, grovelling, or ignoble. He was almost passionless. All that needed for its apprehension more than the pure intellect, or required the exercise of fancy,

imagination, affection, or faith, was distasteful to Cavendish. An intellectual head thinking, a pair of wonderfully acute eyes observing, and a pair of very skillful hands experimenting or recording, are all that I realize in reading his [writings]. His brain seems to have been but a calculating engine; his eyes inlets of vision, not fountains of tears; his hands instruments of manipulation which never trembled with emotion, or were clasped together in adoration, thanksgiving, or despair; his heart only an anatomical organ, necessary for the circulation of the blood. Yet, if such a being, who reversed the maxim "nihil humani me alienum puto" ["I judge nothing human to be alien to me"] cannot be loved, as little can he be abhorred or despised. He was, in spite of the atrophy or non development of many of the faculties which are found in those in whom the "elements are kindly mixed," as truly a genius as the mere poets, painters, and musicians, with small intellects and hearts and large imaginations, to whom the world is so willing to bend the knee.[4]

That genius lay in his particular vision of the world, and his role as a scientist in it. Wilson continued, "His Theory of the Universe seems to have been, that it consisted solely of a multitude of objects which could be weighed, numbered, and measured; and the vocation to which he considered himself called was, to weigh, number, and measure as many of those objects as his allotted three-score years and ten would permit."

Cavendish used a small portion of his principal residence, at Clapham near London, as a bedroom, with the rest crammed full of scientific equipment—thermometers, gauges, measuring instruments, astronomical devices—and tools for making equipment. He turned the upper floors into an astronomical observatory, and the largest tree in the garden quite literally supported Cavendish's meteorological observations. Cavendish was an obsessive instrument rebuilder, and made significant improvements to existing chemical scales, electrical equipment, mercury thermometers, geological tools, and astronomical instruments. But he cared nothing for the outward appearance of his often ungainly looking creations, which have been described by historians of science with phrases like "of rude exterior but singular perfection." (Indeed, his housekeeper was once startled to discover that he had constructed an evaporating device by appropriating the household chamber pots.)

Some historians of science have written of the effect that a scientist's character has on the kind of work that person does. In Cavendish's case, this is surely true, but so is the opposite: Science had an effect on his character. The demands placed on him by the kinds of exacting measurements he undertook surely helped to keep this notoriously neurotic person functional. Not only did these measurements focus his energy constructively, but the respect they gained him among his fellow members of the Royal Society allowed him to maintain what few social connections he had. This respect was merited, for Cavendish's accomplishments were significant and wide ranging. In fact, his accomplishments were even greater than they knew, for Cavendish, who tended to regard his discoveries as his own personal property, did not publish many of them, partly because he was reclusive and partly because he saw them as experimental works in progress, awaiting still greater precision. In a fifty-year career of obsessive work, he wrote fewer than twenty articles and no books. As a result, Ohm's law (which describes the relationship among electrical voltage, resistance, and amperage), and Coulomb's law (which describes the force between two electrically charged bodies) were not named for the man who first came across them. Like masterworks abandoned in the attic by a perpetually dissatisfied artist, these discoveries lay unknown in Cavendish's notebooks for decades, to be unearthed only much later by astonished editors and historians.

Wilson again:

> The Beautiful, the Sublime, and the Spiritual seem to have lain altogether beyond his horizon. . . . Many of our natural philosophers have had a strong and cultivated aesthetical sense; and have taken great delight in one or another or all of the fine arts. For none of these does Cavendish seem to have cared.[5]

Henry Cavendish was drawn, instead, to a deeper, more austere aesthetics. He had an almost instinctive sense for the right kinds of measurements to make and for the simplest way to make them—and then relentlessly drove the precision of his equipment to the edge. His first published work was issued in 1766, when he was thirty-five; it concerned chemical measurements. His last paper, released in 1809, one year before his death, concerned astronomical measurements. In between, he weighed and measured a lot of things very precisely.

One of them was the world. Cavendish's experiment in 1797–98 to determine the density of the world was his masterpiece. It challenged to the fullest even his fanatical quest for precision. He made many other important discoveries, but this one has come to be known as "the Cavendish experiment." Newton's *experimentum crucis* was what historians of science call a discovery experiment, for it revealed a new and unexpected feature of the world in an area where theory was weak. Newton had also plucked it out of a long series of experiments and advanced it as a demonstration standing for all the rest of this work. The Cavendish Experiment, by contrast, was a measurement experiment that stood out by the extreme degree of precision that made it possible at all, was not part of a series, and depended on a relatively well-developed amount of theory. The experiment only gained in importance over time. For while Cavendish used it to measure the density—in effect, the "weight"—of the world, scientists who were putting Newton's law of gravitation into its concise modern form would find that Cavendish's experiment was perfect for measuring the value of the all-important term "G"—the constant of universal gravitation.

THE PATH THAT LED Cavendish to this experiment began, typically for him, with a question about precision—the precision of geographical instruments. In 1763, British astronomer Charles Mason and British surveyor Jeremiah Dixon were sent to the British colonies to settle a long-standing boundary dispute between Pennsylvania and Maryland. The outcome would be the famous Mason-Dixon line, an important boundary in U.S. history in the years leading up to the Civil War. Cavendish wondered how precise their work could possibly be, because the great mass of the Allegheny Mountains to the northwest would exert a very slight gravitational pull on Mason and Dixon's surveying instruments—a pull that was not compensated by an equivalent mass to the southeast, for the water in the Atlantic Ocean is much less dense than stone.

The difference between the densities of mountains and oceans raised, in Cavendish's mind, the question of the average density of the earth itself. This subject was of interest not only to surveyors but to many other kinds of scientists, including physicists, astronomers, and geologists.

According to Newton, the gravitational attraction between two bodies was proportional to their densities. The relative gravitational attraction that astronomical bodies exerted on one another made it possible to know their relative densities; Newton, for instance, had estimated that Jupiter was a quarter the density of the earth. And based on the relative density of matter at the earth's surface and in mines, Newton made an astoundingly accurate guess at its density, writing that "it is likely that the total amount of matter in the earth is about five or six times greater than it would be if the whole earth consisted of water."[6] But nobody had a way to measure it. To do so, one would need to measure the attraction between two objects of known density. The ratio of the attraction between these objects to their densities could be compared to the ratio of the attraction between these objects and the earth to determine the earth's overall density. But the bodies that one could measure in a lab would exert such a tiny gravitational attraction that Newton and others thought it impossible to measure. An alternative would be to measure how much a large land mass of known density (such as a geometrically shaped and geologically uniform mountain) tugged at a small object, such as a plumb bob suspended so that its deviations could be accurately measured. But Newton's calculations made him despair: "[W]hole mountains will not be sufficient to produce any sensible effect," he wrote.[7]

Still, the question of the earth's density was so urgent to astronomers, physicists, geologists, and surveyors that in 1772 the Royal Society appointed a "Committee of Attraction" to attempt a measurement of the density of the earth in what the astronomer Neville Maskelyne described as an effort to make the "universal gravitation of matter palpable." The committee, whose members included Cavendish, decided to try the plumb bob method. In 1775, the Royal Society sponsored an expedition to undertake an experiment—designed largely by Cavendish but carried out by Maskelyne—at a large but regularly shaped Scottish mountain called Schiehallion ("constant storm"). The experiment, predictably, was delayed by bad weather, but after it was concluded, Maskelyne threw such a wild feast for the local Scottish farmers—featuring the consumption of a keg of whiskey and a fire, accidentally started by the revelers, that burnt down the hut the party was held in—that it passed into folklore and was alluded to in a Gaelic ballad.[8]

Back in London, a mathematician calculated from the observations gathered that the density of the earth was 4.5 times that of water, assuming that the ratio of the density of the earth to that of the mountain was ⅖, and that the mountain's density was 2.5 times that of water. Maskelyne was awarded a medal for his measurement, and at its presentation the head of the Royal Society crowed that the Newtonian system was "finished."

Cavendish, naturally, had not participated in the drunken feast, and had not even been on the mountain when the observations were made. Unlike Maskelyne and his Royal Society colleagues, Cavendish was worried about all those assumptions. What made them sure that the ratio of the density of the earth to that of the mountain was ⅖ and that the mountain's density was 2.5 times that of water? Without ascertaining the composition of the mountain and its precise dimensions, the measurement of the earth's density would remain only approximate. A true precision measurement of the density of the earth, Cavendish concluded, would have to be done in the lab, using bodies of known shape and composition. The drawback, as he knew, would be that the force to be measured would be extremely tiny. If the eminent Newton thought that not even a mountain would cause a measurable effect, how could it be done in a laboratory?

In his characteristic style, Cavendish silently mulled over the problem for years while working on other things. Eventually he discussed the problem with one of his few friends, the Reverend John Michell. In addition to being a minister, Michell was a geologist who studied the internal structure of the earth: He had been inducted into the Royal Society in 1760, the same year as Cavendish himself. In 1783, aware that Michell was having health problems while attempting to construct an ambitiously large telescope, Cavendish wrote his friend that "if your health does not allow you to go on with that, I hope it may at least permit the easier and less laborious employment of weighing the world."[9]

Michell, who like Cavendish was also occupied with other experiments, spent a decade building the world-weighing apparatus, but died before he was able to do experiments with it. The equipment ended up in Cavendish's possession, and Cavendish spent a few years rebuilding it to allow for greater precision. He finally began the experiment in the fall of 1797. Although Cavendish was then nearly sixty-seven years old, he applied himself with incredible energy, making observations for hours at a time, obsessively tracking down sources of error, and constantly introducing improvements. His fifty-seven-page paper on the results were published in the Royal Society *Transactions* in June 1798.[10] So much of the article was devoted to Cavendish's fastidious description of his efforts to track down sources of error that one commentator complained that it "reads like a dissertation on errors." It begins simply enough:

> Many years ago, the late Rev. John Michell, of this Society, contrived a method of determining the density of the earth, by rendering sensible the attraction of small quantities of matter; but, as he was engaged in other pursuits, he did not complete the apparatus till a short time before his death, and he did not live to make any experiments with it. . . .
>
> The apparatus is very simple; it consists of a wooden arm, 6 feet long, made so as to unite great strength with little weight. This arm is suspended in an horizontal position, by a slender wire 40 inches long, and to each extremity is hung a leaden ball, about 2 inches in diameter; and the whole is inclosed in a narrow wooden case, to defend it from the wind.

———

Michell intended to measure the attraction between these two-inch metal spheres, placed in barbell-like fashion at each end of the beam suspended from the ceiling, and two eight-inch spheres that would be moved close to the two-inch balls. He would slowly bring the larger weights nearer the smaller ones attached to the beam. That is, if you imagine looking down from the ceiling at the beam and the smaller balls are at, say, 12:00 and 6:00, the larger balls would be positioned at 1:00 and 7:00. The attraction between each pair (one larger, one smaller) of balls would tug the beam, setting it in motion. Because the wire suspending the beam was flexible, the motion would take the form of a tiny back-and-forth oscillation of the beam. Measuring this oscillation would allow Michell to calculate the force of attraction between the balls. This information, together with the known force of attraction between the balls and the earth, contained the information needed to determine the mean density of the earth.

But the second page of Cavendish's paper broaches the key difficulty with this approach: The force of attraction between the balls would be extremely tiny, about one fifty-millionth of their weight. "[I]t is plain," Cavendish wrote, "that a very minute disturbing force will be sufficient to destroy the success of the experiment." The slightest air current, magnetic force, or other extraneous influence would make the experiment impossible. Thus when Michell's equipment came into his possession, "I chose to make the greatest part of it afresh," Cavendish wrote, finding it "not so convenient as could be wished."

"Convenient" was a euphemism. Cavendish worked arduously and unrelentingly to improve its precision. The first thing he did was to make the balls bigger—they became twelve-inch spheres weighing 350 pounds apiece. Even so, it remained essential to guard against disturbing forces—but that, fortunately, was something Cavendish was prepared to go to great lengths to do. The need to reduce and control these forces became the perfect challenge to his obsessive nature.

The most immediate and difficult problem involved temperature differences in the room. If one part of the equipment were even slightly warmer than the surroundings, the air around it would rise, creating currents in the room that would disturb the position of the beam. The body heat of even one person in the room would be completely unacceptable, as would be the heat from a lamp.

> As I was convinced of the necessity of guarding against this source of error, I resolved to place the apparatus in a room which should remain constantly shut, and to observe the motion of the arm from without, by means of a telescope; and to suspend the leaden weights in such manner, that I could move them without entering the room.

Cavendish therefore installed Michell's revamped box-and-ball device into a sealed room in a small building in his garden at Clapham. But to allow the experiment to be operated without entering the room, the equipment had to be redesigned still further. Cavendish remounted the larger pair of weights on a pulley system so that they could be swung around, gradually and slowly, from the outside (Figure 5.1). He attached ivory pointers, using a so-called Vernier Scale, to each end of the barbell, which could determine its position to less than a hundredth of an inch, and installed telescopes in the walls so that these indicators could be read from outside the room. Because he intended to perform the experiment mainly in the dark, he installed a lamp above each telescope with lenses to focus the light through a tiny glass window onto the pointers.

To carry out the experiment, he would slowly swing the larger

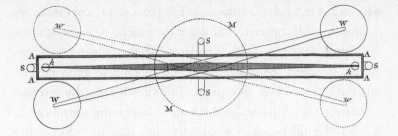

Figure 5.1.   Cavendish's diagram showing a pair of small balls mounted at each end of a beam encased in a box. A pair of larger weights are swung into position nearby.

weights near the case containing the smaller weights mounted on the beam. The attraction between the weights would tug the beam, setting it in motion. Measuring the tiny oscillation that resulted could take up to two and a half hours of careful and steady looking.

In rebuilding this equipment to an extreme degree of precision, Cavendish faced what is known as the "experimenter's trade-off." Each piece has to be just as sturdy and accurate as it has to be, and often no more—for improving it can have undesirable effects elsewhere in the equipment. Making the balls on the beam larger, for instance, would increase the effect but also decrease the precision by straining the beam and the wire supporting it. If he strengthened the beam to compensate, that would place more stress on the wire. Strengthening the wire to compensate would increase the force required to move the rod, reducing the sensitivity of the experiment— and counteracting the effect of the larger balls. Cavendish's genius was to know just how much of a trade-off to make on each piece to maximize both the effect and the precision of the overall experiment.

Although Cavendish's primary worry was air currents, he was also concerned about the contribution of the gravitational attraction of the iron rods used to suspend the large weights and swing them near the smaller ones. This prompted him to remove the weights and measure the attraction of the rods alone, and then to re-place them with copper rods to see if they might be exerting mag-netic attraction. Cavendish wondered whether the wire he was using to suspend the beam was elastic enough; he experimented with the wire, and even though the results implied that the wire was sufficiently elastic, he replaced it with another more suitable wire. Worried that the two-inch metal balls might have picked up a tiny amount of magnetism from being oriented the same way in the earth's magnetic field for too long, he rotated the balls to counteract the effect, and then replaced the balls with magnets to measure what the attraction would be if magnetism were in fact involved. This is an example of what is known as "experimenter's vigilance": If you suspect the presence of a disturbing effect in your experiment, mag-nify it enough to measure it so that you can compensate for it care-fully. Cavendish wondered about the gravitational attraction among the mahogany case surrounding the beam, the smaller balls, and the larger balls, and even though his measurements showed the attrac-tion to be negligible he devoted an entire appendix of his paper to it. All this required more than fastidiousness. To assess what was hap-pening in each case, and to be able to track down, measure, and compensate for any affecting force, Cavendish had to draw on his vast mastery of the scientific knowledge of the day, from electricity and magnetism to heat conduction, mathematics, and gravitation.

Cavendish knew the density of the two pairs of balls and the force of attraction between the balls and the earth. Once he estab-lished the force of attraction between the two pairs of balls, he could use the ratio of the attraction between these objects to their densities to determine the overall density of the earth. "By means of the exper-iments," Cavendish concluded, "the density of the earth comes out 5.48 times greater than that of water"—a result that, he adds with ev-ident satisfaction, is determined "to great exactness." With some glee, he pointed out a discrepancy between this result and the one made twenty-five years earlier with so much fanfare at Schiehallion, one "which differs rather more from the preceding determination than I should have expected." However, Cavendish added with characteristic care and modesty that he would refrain from judgment "till I have examined more carefully how much the preceding determination is affected by irregularities whose quantity I cannot measure."

EARLIER IN THE PAPER, Cavendish had mentioned that one potential source of air currents created "a defect which I intend to rectify in some future experiments." Evidently, he viewed the entire experiment as a work in progress, a momentary respite in a quest for ever more accuracy. He was full of ideas for improvements.

But he would never try the experiment again, though numerous others did. For the next century, scientists would reenact this experiment many times, with new techniques, always seeking greater accuracy—but with negligible improvement. Remarkably, the biggest error in his experiment turned out to be a highly uncharacteristic mathematical mistake, which was spotted by a later scientist.

But a strange thing happened to the experiment over that hundred years: Its purpose evolved. The value of the overall density of the earth grew less important scientifically than the value of a term in an equation that is the modern way of formulating Newton's law of universal gravitation. In modern terms, Newton had said that the gravitational force of attraction F between two spherical bodies of mass M_1 and M_2 separated by a distance r depends on the product of those masses divided by the distance between them squared—multiplied by a constant representing the strength of the gravitational force, known as "G." That is, $F = GM_1M_2/r^2$. Although Cavendish did not know Newton's laws in that form, and the all-important constant G does not appear in his paper, later scientists realized that it is easily measured from his marvelously precise experiment—and the experiment was soon performed for that purpose rather than for the purpose of determining the density of the earth. As one scientist performing the experiment wrote in 1892, "Owing to the universal character of the constant G, it seems to me to be descending from the sublime to the ridiculous to describe the object of this experiment as finding the mass of the earth or the mean density of the earth, or less accurately the weight of the earth."[11]

Almost fifty years after Cavendish's death, in the 1870s, a now-famous laboratory bearing his name was built at Cambridge University, endowed by the chancellor of the university, who was a distant relative of Cavendish.

Today, students still perform Cavendish's experiment with the same basic elements, though with more advanced measuring technologies, such as lasers that bounce off mirrors attached to the balls on the beam to indicate their deflection. Set up the right way, this experiment discloses the strength of the force that holds all matter— the entire universe—together. From this number, one can figure out the behavior of objects orbiting the earth, the motion of the planets in the solar system, and the motion of the galaxies from the time of the Big Bang onward.

The ever-ambivalent Wilson wrote of his biographical subject:

He was one of the unthanked benefactors of his race, who was patiently teaching and serving mankind, whilst they were shrinking from his coldness, or mocking his peculiarities. He could not sing for them a sweet song, or create a "thing of beauty" which should be "a joy for ever," or touch their hearts, or fire their spirits, or deepen their reverence or their fervour. He was not a Poet, a Priest, or a Prophet, but only a cold, clear Intelligence, raying down pure white light, which brightened everything on which it fell, but warmed nothing—a Star of at least the second, if not of the first magnitude, in the Intellectual Firmament.[12]

The beauty that Henry Cavendish created was of a different order. The instrument he used was ungainly, the process tedious, and the mathematics complex. Nevertheless, by virtue of his relentless methodological severity, the way he kept paring back sources of error and replacing inessential pieces until his quarry at last came into view, the Cavendish experiment stands alone in its methodical, stripped, austere beauty.

INTEGRATING SCIENCE AND
POPULAR CULTURE

IN TRUTH, THE GAELIC BALLAD that mentions Maskelyne, *A Bhan Lunnainneach Bhuidhe,* has nothing to do with science, nor with the experiment at Schiehallion to measure the earth's density, nor even with the party afterward. The ballad, written by the fiddler at that party, is about the fiddler's violin ("my wealth and my darling"), which was burnt up in the blaze set off by the revelers. Maskelyne appears only because he fulfilled a promise to replace the fiddle.

In short, the ballad is yet another illustration of science cordoned off from popular culture. In movies, for instance, science generally appears as a pretext for something else—a chase, a whodunit, or good-versus-evil conflict—with the science itself receding into the background. Characters who are scientists appear in a narrow range of superficial roles, such as the smart but evil villain and the nerd—a brainy person in a technical field but otherwise incompetent, unfriendly, and socially inept. In the movies *E.T.* and *Splash!,* the cold, unfeeling scientists nearly kill the vital but somehow also defenseless protagonists.

Much is at stake in how art and popular culture integrate science and scientific issues, for these are important forums in which society processes its ambitions and anxieties. The general lack of success of art and popular culture to integrate well science and scientific issues is thus disturbing, given how fundamentally and inextricably science is woven into contemporary life, and has been ever since the time of Galileo. The constantly recycled stereotypes of science as cold and distant makes it seem remote, impersonal—hence threatening and potentially dangerous. It also undercuts any attempt to foster the appreciation of beauty in science, for it prevents us from seeing how deeply science permeates the world and its wonders.

It is difficult even for well-intentioned artists to integrate science into their work; if you find that appreciating the beauty of scientific experiments requires preparation, you should try certain works of science-inspired art. Many examples were on display in 2003, the fiftieth anniversary of the discovery of the structure of DNA, when several art exhibitions were devoted to the subject. These prompted *New York Times* reviewer Sarah Boxer to quip that, "Like DNA, DNA art demands decoding." She likened her experience at some galleries to hearing a musical performance and having someone constantly whisper in your ear what each phrase means. "[I]f you want to understand the DNA connections," Boxer wrote, "you are in for a lot of reading."[1]

Theater is a good forum for integrating science, given the complex human situations it is able to depict. Even there, however, historical and scientific truths are frequently altered to make the human situation plausible, or even presentable. An example is Heinar Kipphardt's play *In the Matter of J. Robert Oppenheimer*. It was based on transcripts of Oppenheimer's famous hearing in 1954, when the man most responsible for the success of the Manhattan Project had his attempt to restore his security clearance rejected, largely because he had antagonized enemies by his initial opposition to the hydrogen bomb, but also because of his past left-wing politics. But Kipphardt found it necessary to invent a fictitious final speech for Oppenheimer and to make other adjustments, to all of which Oppenheimer vociferously objected. (Oppenheimer worked with a French actor-director to modify Kipphardt's work, but the more historically accurate result was lifeless.)

Playwrights whose work is scientifically sensitive—meaning that it effectively utilizes scientific terms, imagery, and ideas for theatrical effect—include Tom Stoppard, in plays such as *Hapgood* and *Arcadia*. Scientists whose work is theatrically sensitive—meaning that it effectively uses theater to dramatize scientific issues—include Carl Djerassi in his play *The Immaculate Misconception*, and Djerassi and Roald Hoffmann in their play *Oxygen*.

One of the few plays to fully and successfully integrate science and theater is Michael Frayn's *Copenhagen*, whose three protagonists are the two scientists Werner Heisenberg and Niels Bohr and Bohr's wife, Margarethe. Much has been made of the fact that this play incorporates some audience members into the proscenium—they sat in tribunal-like stands in the back of the stage, facing the audience—suggesting that every observer is observed and no observer can observe himself, in a theatrical equivalent of the uncertainty principle. This is not, I think, the best way to state the originality of the play, for the self-consciousness of theatrical actions as a produced artifact is as old as theater. More remarkable, and the key to the successful integration of science in this play, is the role of Margarethe. She serves in a sense the function of the ancient chorus; she is our stand-in. But she is not just an observer, an interested and sympathetic layperson who demands to be kept informed in ordinary language. She is implicated in the very events that she is involved in; she even literally helped compose some of them. Niels Bohr found writing extremely difficult—to the extent that some people have wondered whether he was dyslexic—and most of his work and correspondence were dictated to others, including Margarethe. She was therefore at least in part an accomplice and has no illusion of being able to stand back from the events that she is questioning.

Margarethe thus serves to remind us how implicated we already are in science. Some approach science as if it were no more than a giant corporation nested in the social world. But science is so tightly woven into contemporary society—so integral to how we understand ourselves and our relation to the world—that it is impossible to get that kind of distance from it. Science is less like a corporation and more, as it were, like the entire commerce system, any modification of which will ripple through human society in myriad, often unpredictable ways. This intimate and inextricable interweaving of science and human society and self-understanding suggests that *Copenhagen* need not be the exception—that hundreds of other *Copenhagen*-like plays could and should be written. It also implies the presence of more science than we think in the beauties that we already have.

Thomas Young's diagram of an interference pattern (bottom) produced by passing light (top) through two closely spaced slits.

Six

LIGHT A WAVE:
Young's Lucid Analogy

—

THE ENGLISHMAN THOMAS YOUNG (1773–1829) RECEIVED
a strict upbringing as a member of the Society of Friends (Quakers). By the time he turned twenty-one, he had given up actively
practicing that religion, and derived much pleasure from participating in music, the arts, horseback riding, and dancing. But much
of his character was formed by his Quaker background, which
gave him both strengths and weaknesses. As per the Quaker ideal,
he was straightforward, courteous, and direct, with an independent
and persevering mind. These traits certainly helped him uncover
the wave (or "undulatory," in the language of the day) nature of
light—which fundamentally challenged the prevailing particle (or
"corpuscular") view promoted by Newton. But Young also had the
Quaker propensity to be laconic to the point of appearing cold. He
could frustrate people by confidently proclaiming a terse conclusion
without bothering to reveal the reasoning behind it. At times this
hampered his career and the reception of his work.

At the same time, his penchant for directness and economy was
reflected in his ability to devise demonstrations whose point was

plain and all but irrefutable. The most famous was his two-slit experiment—now often called simply "Young's experiment"—a dazzlingly simple proof that light behaved, despite Newton's belief, as if it were a wave, not a stream of tiny particles. Young's experiment is a classic example of the successful use of analogy in science. By lucidly displaying light acting wavelike, it produced an "ontological flash"—the disclosure of new meaning as things appear fundamentally different from how they had previously been perceived.[1]

YOUNG WAS RECOGNIZED as a prodigy soon after his birth in 1773. By age two he had learned to read; by age six, he had read the Bible cover to cover—twice—and had begun to teach himself Latin. He soon mastered more than a dozen languages. He was one of the first to decipher Egyptian hieroglyphics, and would play a key role in decoding the Rosetta Stone.[2]

From 1792 to 1799, Young studied medicine but was ultimately unsuccessful as a practitioner, partly because of his lack of a comforting bedside manner. During this time, Young became interested in vision and especially in the structure of the human eye, an extraordinarily adaptable and complex lens. Other medical studies got him interested in sound and the human voice, and he began to wonder whether sound and light were fundamentally similar. Sound was known to be created by waves in the air, and Young became convinced that light, too, consisted of waves. This challenged the prevailing theory that light was made of minute particles—Newton's term was "corpuscles"—that traveled in straight lines from their source to the eye.

A few wavelike aspects of light had been pointed out by various scientists in the 1660s. One was diffraction: The Italian scientist Francesco Grimaldi had noticed that when light passes through a narrow slit onto a wall, the edges of the narrow band of brightness are slightly blurred, suggesting that light diffracts, or bends slightly, around the edges of the slit. Another was refraction, or the bending of a ray of light as it enters another medium, which Newton's nemesis Robert Hooke pointed out could be explained better if light consisted of waves rather than corpuscles. And the Danish scientist Erasmus Bartholin had discussed the strange phenomenon of double refraction, noticed in a type of crystal found on an expedition to Iceland in 1668. When a ray of light fell on Icelandic spar, as the crystal was known, it split into two rays that behaved differently— a phenomenon that puzzled scientists of the day and seemed difficult to explain by the corpuscular theory.

But these effects were tiny—minute enough to tempt scientists to overlook them—and it was not clear how or even whether they were related to one another. Newton in particular had articulated persuasive counterarguments against the wave view, pointed out many observations in contradiction with that view, and expected that some other explanation might be found for the small anomalies of diffraction and refraction. As Newton wrote in his *Opticks* of 1704, waves do not travel in straight lines but bend around objects that stand in their way—which light does not appear to do.

> The waves on the surface of stagnating water, passing by the sides of a broad obstacle which stops part of them, bend afterwards and dilate themselves gradually into the quiet water behind the obstacle. The waves, pulses or vibrations of the air, wherein sounds consist, bend manifestly though not so much as the waves of water. For a bell or a cannon may be heard beyond a hill which interrupts the sight of the sounding body; and sounds are propagated as readily through crooked pipes as straight ones. But light

is never known to follow crooked passages. . . . For the fixed stars, by the interposition of any of the planets, cease to be seen. . . . [3]

Despite Newton's authority, Young was fascinated by the idea that sound and light were analogous phenomena. Because his medical practice made few demands on his time or his interest, he was able to throw himself into scientific inquiries on the subject. He regularly attended meetings of the Royal Institution, a newly formed organization whose purpose was to diffuse "the knowledge of useful mechanical improvements" and to "teach the application of science to the useful purposes of life," and gave up medicine to join its staff as a professor in 1801. One of his chief duties was to prepare and deliver a series of lectures to the society's membership on "natural philosophy and the mechanical arts." These lectures illustrate Young's strengths as a career scientist. Indeed, they are a goldmine for today's historians, because they summarize accurately and concisely practically the entire spectrum of scientific knowledge of the day; it is hard to think of a branch of science about which Young was not as informed as any specialist. Not only that, he used the lectures to introduce several fundamental concepts; in one, his audience heard the word "energy" used for the first time in its modern scientific sense. Nonetheless, the lectures must have been a trial to attend, because Young's clipped, abbreviated style, coupled with the wide-ranging subject matter, made them an exhausting, speedy intellectual tour de force. Young, in fact, lasted only two years as a professor at the Royal Institution. The Royal Society found a much greater match for his talents in 1802, when it appointed him its Foreign Secretary. He held that post, which took advantage of his command of languages, for the rest of his life.

But the year before he joined the Royal Institution, in 1800, Young published his first major work exploring the analogy between sound and light, the "Outlines of Experiments and Inquiries Respecting Sound and Light."[4] It would take Young several years to come up with the experiment, bearing his name, that would clinch the analogy. But the paper of 1800 was an important first step, and a milestone in scientific literature, for it described for the first time the concept of interference on which his famous experiment would be based: the way that, when two waves cross, the resulting motions combine the effects of the motions of each wave separately. "Interference" is an unfortunate name for this phenomenon, since it suggests something illegitimate, corrupt, or degraded, when what is happening is that two things are combining to create something new. Perhaps recognizing this, Young sometimes used the more elegant term "coalescence."

Newton had partly anticipated the idea of interference when he explained the tides at Batsha, a port of the Kingdom of Tongkin near modern-day Haiphong in Vietnam. British merchants of the seventeenth century, seeking trade with Tongkin, knew its coastal waters to be unusual. In 1684, an English traveler who had spent time in Batsha published a letter in the *Philosophical Transactions* describing the curious tidal pattern: Every fourteen days, there was no tide at all—the water level neither rose nor fell that day—and in between only a single tide, which slowly rose to a peak after seven days and then subsided. This queer behavior attracted the interest of scientists, and Newton proposed an explanation in his masterwork, the *Principia* (1688). The ocean tides, he said, were reaching the port from two different seas—the China Sea and the Indian Ocean—via two different channels of different lengths, which made one arrive in six hours and the other in twelve. The combined effect—the high tide from one direction often compensating for the low tide coming from the other—eliminated one tide, and twice each lunar month eliminated both, leaving the water level unchanged.[5] But while this is now conceived as an example of wave interference, Newton did not generalize the insight and conceive it as a property of waves but saw it instead as the effect of a superposition of particular motions that only occurred at one special place.

Young's 1800 paper only discusses the concept of interference in connection with sound waves, and does not yet explicitly generalize it to light even though much of the paper is about light. Nevertheless, Young's insight was to identify interference, to realize that it was a basic feature of wave motion, and to understand that it happened simultaneously at many locations wherever waves cross. However, the description he gave obscured the originality of the concept and even his role in its invention. Young did not draw attention to the concept—he spoke merely of the fact that when sound waves cross one another each particle of the medium (water or air molecules, say) partakes of both motions. He claims no priority for its discovery, makes it sound obvious and well understood, and modestly introduces it in correcting the work of another scientist.[6]

The next year, Young extended the concept of interference to water and to light. He would write later:

> It was in May 1801 that I discovered, by reflecting on the beautiful experiments of Newton, a law which appears to me to account for a greater variety of interesting phenomena than any other optical principle that has yet been made known.
>
> I shall endeavour to explain this law by a comparison. Suppose a number of equal waves of water to move upon the surface of a stagnant lake, with a certain constant velocity, and to enter a narrow channel leading out of the lake. Suppose then another similar cause to have excited another equal series of waves, which arrive at the same channel, with the same velocity, and at the same time with the first. Neither series of waves will destroy the other, but

their effects will be combined: if they enter the channel in such a manner that the elevations of one series coincide with those of the other, they must together produce a series of greater joint elevations; but if the elevations of one series are so situated as to correspond to the depressions of the other, they must exactly fill up that depression, and the surface of the water must remain smooth; at least I can discover no alternative, either from theory or from experiment.

Now I maintain that similar effects take place whenever two portions of light are thus mixed, and this I call the general law of the interference of light.[7]

In the interference of water waves, the elevations—the technical term is "amplitude"—of different waves can combine to reinforce one another, forming spots of even greater elevation, while in "destructive interference" the elevations and depressions of different waves can combine to leave the water surface unchanged. Something similar happens in the case of light interference, where the amplitude of a light wave is connected with its intensity. Wherever the amplitudes of interfering light waves combine to reinforce one another, they form spots of greater light intensity; wherever the amplitudes are in opposed directions, they cancel one another out and form dark spots.

Young put the concept of interference to work shedding light on many puzzling phenomena. The most dramatic of these was his explanation of Newton's rings, the series of concentric bands that appear when a convex lens is pressed against a glass plate. Young extended Newton's own account of these rings by showing that the dark areas of these rings were products of destructive interference.

And while Young's expositions were sometimes obscure, his demonstrations were not; they were clear and simple, and grew out of his thorough understanding of the subject. In 1803, for instance, he read a paper before the Royal Society, called "Experiments and Calculations Relative to Physical Optics," which began as follows:

> In making some experiments on the fringes of colors accompanying shadows, I have found so simple and so demonstrative a proof of the general law of the interference of two portions of light . . . that I think it right to lay before the Royal Society a short statement of the facts which appear to me so decisive. . . . [T]he experiments I am about to relate . . . may be repeated with great ease whenever the sun shines, and without any other apparatus than is at hand to every one.[8]

In the first of these experiments, Young used a needle to puncture a tiny hole in the thick paper he had used to cover a window, making a thin shaft of light fall on the opposite wall. When he inserted "a slip of card about one-thirtieth of an inch in breadth" into this sunbeam, it created a small shadow with colored fringes not only on either side of the shadow, but also diffracted into the shadow itself. In this shadow, he observed the series of parallel black-and-white bands that is now known as the distinctive signature of an interference pattern.

In his Royal Institution lectures, published in 1807, his diagrams and demonstrations were spectacular. His twenty-third lecture, "On the Theory of Hydraulics," applies the concept of interference to water waves. To accompany it, he built a shallow tank with two sources of waves. The crests and troughs of the two sets of waves give rise to a stationary pattern that makes the interference pattern clearly visible. The device was the prototype of the ripple tank familiar to most high school physics students (Figure 6.1).

And in his thirty-ninth lecture, "On the Nature of Light," Young applies the concept of interference to light. To accompany this lecture, he came up with a demonstration that is not only the most direct way to illustrate light interference, but also the classic demonstration of light acting as a wave. Young described his demonstration as follows:

> [A] beam of homogeneous light falls on a screen in which there are two very small holes or slits, which may be considered as centers of divergence, from whence the light is diffracted in every direction.

The two holes or slits become, in effect, two wave sources, like the two sources in the ripple tank. And while one looks down on the interference pattern in a ripple tank, seeing two sets of overlapping circles with lines radiating out from a spot between the two sources, the observer of this experiment must watch the pattern as the light falls on a screen.

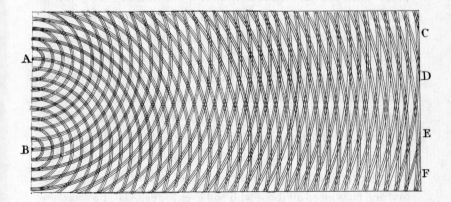

Figure 6.1.   Young's diagram of an interference pattern produced by a series of waves from two sources.

In this case, when the two nearly formed beams are received on a surface placed so as to intercept them, their light is divided by dark stripes into portions nearly equal, but becoming wider as the surface is more remote from the apertures, so as to subtend very nearly equal angles from the apertures at all distances, and wider also in the same proportion as the apertures are closer to each other.[9]

The interference pattern now consists of parallel stripes of light, with the bright bands being areas where the light waves reinforce one another, and the dark bands areas where they act destructively on one another.

I've read popular-science books that claim this experiment can be easily performed at home—all one needs is a flashlight, a pin, some pieces of cardboard, and a dark room. Don't believe it; I wasted an entire afternoon trying. It can be done, but this experiment requires great care to work right. It is easy not to see the bands or to see at most shadows that are produced by diffraction—by the light bending around the edges of the cardboard, or around imperfections in the holes you have made if you are the slightest bit clumsy. It is possible to do with paper, cardboard, and a razor blade, but the sharpness of the edge is important—some science-education supply companies make plastic squares with slits on them for this purpose. But doing it right is so difficult that science historian Nahum Kipnis read Young's lectures carefully enough to peer past Young's disarmingly simple Quaker prose to realize that even he misinterpreted, at least once, a diffraction pattern as an interference pattern.[10]

It would be nice to say that Young's demonstration was a landmark event in the triumph of the wave theory over the particle theory, and convinced all those who had eyes to see. Alas, this was not the case, for several reasons.

A first was, again, Young's style. Although his measurements were precise and his calculations mathematically adept, he rarely bothered to explain his reasoning, record his actual measurements, or even provide extensive descriptions of his experiments. This seems to have stopped many colleagues from understanding him, or at least prevented them from being persuaded. Furthermore, the modest Young usually went out of his way to avoid claiming originality for the wave theory of light and the concept of interference. At one point, in 1801, giving voice to an overcharitable reading of Newton's writings, Young even claimed that his predecessor "was in reality the first that suggested such a theory as I shall endeavor to maintain." This, too, dampened appreciation for the originality of his ideas.

A second reason was that Young had the misfortune to become the target of Henry Brougham (pronounced "broom"), an influential correspondent who wrote for the *Edinburgh Review*, a trendy new literary magazine. Brougham worshipped Newton and blasted Young, who dared disagree with the master, in a three-part and anonymous vitriolic attack. A sample:

> We demand, if the world of science which Newton once illuminated, is to be as changeable in its modes as the world of fashion, which is directed by the nod of a silly woman or a pampered fop? Has the Royal Society degraded its publications into bulletins of new and fashionable theories for the ladies of the Royal Institution? *Proh pudor!* [What a disgrace!] Let the Professor continue to amuse his audience with an endless variety of such harmless trifles, but in the name of science, let them not find admittance into that venerable repository which contains the works of Newton, and Boyle, and Cavendish, and Maskelyne, and Herschel.[11]

The usually unflappable Young was angered, and he replied, as one often did in the nineteenth century, by writing a pamphlet. But scientists are usually not well equipped to carry on this kind of fight in public—they are trained to convince other scientists, not the public—and Young's reply, written in a dry, annoyed style, was much less scintillating than the attack. Full of dull, true, but defensive-sounding claims like "Let him make the experiment, and then deny the result if he can," Young's reply sold exactly one copy.

With Young less than adept at promoting his ideas, the wave theory of light spread slowly. Some fifteen years after Young's demonstration, French scientist Augustin Fresnel rediscovered the phenomenon of interference, developing a variation of the two-slit experiment in which the beam of light is separated into two sources by a flat prism, now called a "Fresnel biprism." (Ever since, as we shall see again in Chapter Ten, Young's experiment has been done in two classic variations—using Young's double-slit method and using Fresnel's biprism.) French scientists' enthusiasm for this finding led the mainstream scientific community finally to accept the wave theory of light—and belatedly to give Young his due.

The phenomenon of interference not only established the wave theory of light, but provided a useful tool for scientific inquiry, since an interference pattern is simple and easily recognized. If a phenomenon can be made to exhibit an interference pattern, that thing is wavelike.

Wave theory still had problems, most scientists thought. Particularly vexing was the question of the medium in which the light waves traveled. Sound waves were waves of air, just as water waves were waves of water. What was the analogous medium for light? What was light a wave *of*? The traditional answer was an invisible substance called the "ether" that supposedly permeated all of space. When the human eye saw a star, it was responding to a wave in the ether that had begun at the star and rippled through space in the ether until it washed up against the retina. Then, at the end of the nineteenth century, Albert Michelson and Edward Morley demonstrated that the interference pattern of light beams that had traveled in two different directions could be used to measure how fast they had traveled with respect to each other. Their inability to detect any difference was taken as a sign that the ether did not exist, that light waves somehow did not need to travel in a medium. Their experiment did not transform our understanding of light—which was still understood to be wavelike—so much as it did our understanding of waves. The Michelson-Morley experiment would soon become an important piece of evidence for Einstein's theory of relativity.

In the nineteenth century, Young's two-slit experiment, extending the wave analogy from acoustics to light, heralded a paradigm shift from the particulate view of light to the wave view. In the twentieth century, a still more dramatic extension of Young's experiment would be enacted—a third double-slit experiment involving not water waves or light waves but particles. This further application of the wave analogy would be the most dramatic simple demonstration of the mystery of quantum mechanics—the final experiment in this book, and the one many scientists consider the most beautiful of all.

SCIENCE AND METAPHOR

YOUNG'S EXPERIMENT, whose beauty stems from the clarity with which he made one phenomenon—light—show itself acting like another—waves—is a classic example of the successful use of analogy in science. But analogy (from the Greek for "proportionate") and metaphor (a figure of speech by which one thing is spoken of as if it actually were another, from the Greek for "carrying further") can also mislead and interfere with thought. For this reason scientists are of two minds, so to speak, about their value.[1]

Some find analogies and metaphors distracting at best, confusing at worst. "When thinking about nature," the biologist Richard Lewontin once wrote, "beware of metaphors." Physicist Ernst Mach thought it useful to be able to say that "the fact A now considered comports itself . . . like an old and well-known fact B," but denied such assertions played a structural role in science. Like Mach, Pierre Duhem, a physicist and historian of science, regarded metaphors and analogies as important psychological, explanatory, and educational tools. But, Duhem insisted, real science eventually discards them.

Metaphor-users and analogists, in this view, play Aaron to Moses. Just as the prophet Moses wrested knowledge from the beyond, which his brother and spokesman Aaron, a careful listener who dwelt more intimately in the community, then transmitted to the masses, so scientists discover truths about nature, which educators, popularizers, and journalists interpret for novices and the public using imagery and everyday language. The anti-metaphorists and anti-analogists view them as having a place in this secondary process involving the dissemination and transmission of information, but not in the primary discovery process itself. Science, real science, is about what something *is*, not what it's like.

Others, however, find metaphors and analogies to be so deeply woven into scientific thought as to be practically indispensable. "It is probably no exaggeration to say," remarks physicist Jeremy Bernstein, "that all of theoretical physics proceeds by analogy." Writes physicist John Ziman, "We cannot think about *anything*, except by analogy and metaphor." The pro-metaphorists and pro-analogists claim that whenever scientists say what something is, they are also, inevitably, saying what it's like and what's like it.

This kind of conflict—where opposing armies mass on either side of a seemingly clear and intractable boundary—can be resolved by philosophy, whose role is to detect and expose the confusions and ambiguities that make such boundaries seem intractable in the first place. In the case of metaphors in science, a philosopher would instinctively point out that not all metaphors work in the same way, or for the same reason. Metaphors in science, in fact, work in at least three different ways.

A first use of metaphor is as a filter. Consider classic metaphors such as "man is a wolf" or "love is a rose." In these, what is known as the "secondary subject"—wolf, rose—calls the reader's attention to certain conventionally understood features of these subjects (solitary and predatory in the case of the wolf; pretty but thorny for the rose). The goal in both cases is to illuminate those aspects in the "primary subject"—man or love—and filter out, as it were, the rest. Filtrative metaphors allow a first grip on the primary subject. But because all filters omit things, they can be deceptive if taken too literally. Lewontin's ire, for instance, was sparked by a colleague's reference to DNA as a genetic "program" on the basis of which, it was then claimed, one could "compute" an organism. An organism, he rightly observed, is *not* a computer. But the question was not whether this filtering metaphor was entirely true in each and every sense, but whether it provided a quick, illustrative flash of intuitive understanding of some aspect of the subject. The point is to keep moving forward, and danger lies in becoming stuck by taking a metaphor too literally.

A second use of metaphors is creative. Here the priority of the two terms is reversed, for the secondary subject is used to bring to bear an already-organized set of equations on the primary subject—and it grows into the salient, technically correct term, often expanding its meaning in the process, with the primary subject being merely one of its derivative forms. When Young and others began to call light (the primary subject) a wave (the secondary subject), it was an example of this kind of analogy. Waves originally referred to a state of disturbance propagated from one set of particles in a medium to another set, as in sound and water waves. When Young and others began calling light a wave, they assumed that it, too, moved in some medium but had no real sense of what this medium (called, by default, the "ether") might be. By the end of the nineteenth century, scientists had begun to think that light propagated in the absence of a medium—but the same equations governed light, which it was still correct to describe as a wave. "Wave" became the technically correct term, and was transformed in the process, because scientists' understanding of the concept of "wave" changed when they applied it to a disturbance that can be propagated without a medium (it changed still further when waves showed up in quantum mechanics). This is the kind of analogous extension that Bernstein noted was the basic process of theoretical physics: seeking to understand the unfamiliar by comparing it to things we know, and adapting our descriptions of the known in the process. We often discover what something *is* by discovering what it is like. The meanings of our old terms then change to become "like" the new. "The literal," a science historian once insisted to me, "is just a metaphor for metaphor." Or, to paraphrase an old saying: one who analogizes invents. Philosopher Eugene Gendlin calls this "just-as-ing," an active process in which something new emerges out of transformation of the old, rather than a process in which something old is imposed on something new.

Another example of this creative use of analogy in science involves the concept of energy. In the beginning, the subjective experience individuals have of themselves as a center of action was one factor.[2] And in his eighth Royal Institution lecture—"On Collisions"—Young said that, "The term energy may be applied, with great propriety, to the product of the mass or weight of a body, into the square of the number expressing its velocity," an expression that would be written today as mv^2. Young thereby used the word "energy," apparently for the first time, in its modern sense. But Young's "energy" was not strictly ours, only what we call "kinetic energy," and not even our formulation of it (we say it's actually $\frac{1}{2} mv^2$, not mv^2). This evolution would extend through the rest of the century.

A third use of metaphor seeks to recast our overall view of something.[3] One example is the physician Lewis Thomas's well-known remark that the earth is not like an organism, but "*most* like a single cell." Another example of a recasting metaphor is the late biologist Stephen Jay Gould's thought experiment of "replaying life's tape," in which he tries to restructure our perception of evolution as a ladder of progress or cone of increasing diversity into an appreciation of its contingency. "You press the rewind button and, making sure you thoroughly erase everything that actually happened, go back to any time and place in the past . . . let the tape run again, and see if the repetition looked at all like the original."[4]

These three ways of using metaphors in science are not hard-and-fast, and one finds uses that blend two or more of them. Still, noting them does much to explain why scientists can trip themselves up on the subject, saying different things about it and appearing to contradict themselves.

Clarifying the nature of metaphoric usage is important for understanding science and its beauty. One reason is that the secondary subjects of metaphors are rooted in culture and history. Scientists always work with culturally and historically transmitted concepts and practices. It is always transforming, not transcending, what it has been given by culture and history.[5]

Metaphors and analogies are focused ways in which human beings apply everything they have inherited and developed to project themselves into the future. Training and experience fill our minds with metaphors, which we cannot avoid bringing to bear on the new, transforming what we know in the process. Thus it can be both true, as Ziman says, that we can't think without metaphors, and that, as Peter Galison says in his 1997 book *Image and Logic*, which struggles with analogies for the poorly understood relation among theory, experiment, and instruments, "All metaphors come to an end."

The metaphor user, therefore, cannot be seen as playing Aaron the listener and transmitter to Moses the prophet and discoverer. Or, if one insists on putting it that way, one must recognize that the difference between prophecy and listening, between primary discovery and secondary transmission, and between saying what something *is* and what it's *like* breaks down. Every act of research is already metaphorical thinking. For Moses, one might say, played Aaron to God.

Foucault's pendulum at the Panthéon.

Seven

SEEING THE EARTH ROTATE:
Foucault's Sublime Pendulum

—

THE FIRST FOUCAULT'S PENDULUM I EVER SAW WAS AT the Franklin Institute in Philadelphia, the city where I was born. The pendulum hung—and still hangs—in the well of a main stairway. Its thin wire cable was attached to the ceiling four stories up, while its silver bob glided silently back and forth over a compass dial (which has recently been replaced by a backlit globe) embedded in the floor. I can still recite the information on the first-floor sign: The pendulum's cable was eighty-five feet long and its bob—a steel sphere twenty-three inches in diameter and loaded with lead shot—weighed 1,800 pounds. The bob swung back and forth in a straight line, silently and ponderously, once every ten seconds. The plane of its swing slowly shifted to the left (clockwise) at an unchanging rate throughout the day: 9.6 degrees per hour. The sign informed me that although the pendulum seemed to be changing direction, this was false; the pendulum always swung in exactly the same direction with respect to the stars. Instead, what the museum visitor was really seeing was the earth—and along with it the floor of the Institute building and the compass dial in that floor—turning underneath the pendulum.

The pendulum had been installed in 1934, when the Institute moved to its current building. Its installation was cause for an unusual parade. The wire, which weighed only nine pounds, could not be rolled up but had to be kept straight to prevent kinks and stresses that would interfere with its swing. It was therefore carried, fully stretched out, through the streets of Philadelphia from the manufacturer to the new building. The slow and bizarre procession of eleven men carrying a long wire was accompanied by a police escort and trailed by bemused onlookers and reporters.[1]

The Franklin Institute's pendulum signaled its change of direction by knocking down, every twenty minutes or so, one of the set of four-inch steel pegs that stood in two semicircles on the floor, tracing the outside of the dial. Whenever I visited the Institute, I would often pull myself away from the exhibit I was playing with to rush back and join the crowd of onlookers watching the swing of the silver bob and staring at the pegs, hoping to see one fall. First the bob would graze a peg, making it shiver. A few swings later, the peg would actively wobble. A few more and the tip of the bob would strike the peg solidly enough to make it rock back and forth. Not long now! One or two more swings and the peg would go over—*plink!*—and the bob would begin to creep toward the next peg. Sometimes I'd just stare at the pendulum itself, trying to obey the sign and make myself see that it was *I*—and the solid floor beneath my feet—that was moving. For reasons I did not understand, I never quite succeeded, though the pendulum did leave me with a feeling of mystery and awe.

The pendulum's movement was entirely beyond my control, as inexorable a performance as I knew, then or now. The only human influence on it was the museum staff member who started it swinging in the morning, in the north-south direction, just before the museum opened at 10:00 A.M. Sometimes I would arrive at the museum early and wait for the doors to open, so that I could race to the stairwell to try to catch the pendulum being started. I was always too late. Once I heard that a museum supporter had arranged, as a birthday gift, for his son to start the pendulum one day. How I envied that child! Other kids may have dreamed of tossing out the first pitch at a baseball game. I dreamed of starting a Foucault's pendulum.

FRENCH SCIENTIST Jean-Bernard-Léon Foucault (1819–1868) was born in Paris. As a youth he built scientific and mechanical toys, and began studying medicine with the aim of exploiting his practical talents by becoming a surgeon—until he discovered his aversion to blood and suffering. His interest reverted to mechanical instruments and inventions, and he became fascinated by the new photographic imaging processes developed by his fellow Frenchman Louis Daguerre. In work that successfully tapped his mechanical abilities, Foucault teamed up with another ex–medical student, Hippolyte Fizeau, to improve what were known as daguerreotypes, which were predecessors of the modern photograph. The two took the first clear picture of the sun in 1845, and then—first working together, then separately after a personal dispute—showed that the speed of light was greater in air than in water, in 1850, before going on to measure the absolute speed of light in air. Still later, Foucault made significant contributions to the construction of mirrors for telescopes.

Foucault also took some of the earliest photographs of stars, a technical tour de force at the time. Normally one would photograph faint objects by opening the camera shutter for minutes at a time. But because the earth rotates on its axis, the stars seem to slowly move in the sky, making it impossible to simply leave the shutter open. Instead Foucault, reviving a once-discarded idea, built a pendulum-driven clockwork device that would keep the camera pointed at a star long enough for an exposure—though in place of a bob on a string he used a metal rod that vibrated like a pendulum when twanged. (I have read dozens of articles on pendulums, and spoken to many scientists about them, and I assure you that the technical term for starting a metal-rod pendulum is to "twang" it.)

Much of this work took place in a laboratory Foucault set up at his home on the rue Assas in Paris. One day he put a rod in his lathe, mounting it on a chuck or bit that could spin freely, the way a skateboard wheel can spin freely on its mount. When he twanged the rod and turned the lathe slowly, he was startled to see the rod continue to vibrate back and forth in the same plane. Curious, he experimented with a more conventional pendulum—a spherical weight suspended vertically with piano wire that could swing freely. He attached it to the mount of a drill press and turned the bit. The pendulum, too, continued to swing in the same plane.

If one stops to think about it, this is not surprising. According to Newton's laws, a body in free motion, such as a pendulum bob, moves in the same direction unless some force is applied to change it. Because turning the freely spinning bit did not apply any force to the rod or pendulum, they continued to oscillate in the same direction. But the unsurprising can still be unexpected. Foucault soon realized that this effect, if magnified enough, could be used to demonstrate the diurnal (daily) rotation of the earth on its axis.

Later he summarized the reasoning process rather elegantly as follows. Imagine we build a little pendulum on a table atop a freely and smoothly swiveling plate (we might say a lazy Susan). We have now what Foucault called a *petit théâtre* on which we are about to stage a performance. The lazy Susan is like the earth and the surrounding room like the rest of the universe. If we set the pendulum swinging in a plane—let's point it at the door—and then slowly swivel the plate, what happens? At first we might expect that the pendulum's plane of oscillation would turn along with the base. *Erreur profonde!* The plane of oscillation is not a material thing attached to the plate. Because of the pendulum's inertia, the plane of its swing is independent of the plate—it now "belongs," so to speak, to the space around it rather than to the plate. Whichever way we turn the plate, the pendulum continues to point at the door.

This performance in the little theater demonstrates that the lazy Susan moves, while the pendulum's plane of oscillation is unchanged. But imagine that we make our little theater very big, Foucault says. Imagine, in fact, that we—as well as the rest of the room and everything we can see around us except for the sun, planets, and stars—are on board the turning plate. Now it will look to us like we are motionless and the pendulum's direction of oscillation is changing. Again—*Erreur profonde!* We are the ones who turn. But Foucault points out an additional complication. Our little pendulum is in the center of a flat plate, so that a complete turn would change the pendulum's plane of oscillation by 360 degrees, or a complete circle. But an earth-based pendulum is on the surface of a sphere. Depending on where the pendulum is located between pole and equator, a complete rotation of the sphere will make the pendulum's plane turn by different amounts—and the sphere will have to turn by different amounts to get the pendulum's plane of oscillation to make a complete rotation. Doing the math, Foucault calculated that the number of degrees through which the pendulum's plane of oscillation would shift in twenty-four hours would be 360 degrees times the sine of the latitude—which thus provided a way to determine the person's north-south location on the globe. But the details of the calculation do not matter nearly as much as the visible demonstration of the effects of the earth's rotation.

Foucault wondered whether he could see the effect of the

earth's rotation using a real pendulum. He suspended a pendulum from the vault of his basement, using a thin six-and-a-half-foot wire and an eleven-pound bob. On Friday, January 3, 1851, he tried it for the first time. To make sure the pendulum's swing would be steady and straight, he tied the bob to the wall with a cotton cord, let its motion come to a complete halt, then burnt through the cord with a candle. Although the experiment seemed to work, the wire broke. Five days later, on Wednesday, January 8, 1851, at 2:00 A.M., he got it working again, and within half an hour he found that the "displacement is such that it is evident to the eye," and that "the pendulum turned in the direction of the diurnal movement of the celestial sphere."[2] Ever methodical, however, he found it less interesting to watch the phenomenon on a grand scale and "more interesting to follow the phenomenon more closely, so as to be satisfied of the continuousness of the effect."[3] He mounted a pointer on the floor so that it would just touch a point on the pendulum, and noticed that, in less than a minute, the pendulum was displaced toward the left of the observer—meaning that the pendulum's plane of oscillation was moving with the apparent motion of the heavens.

A few weeks later, Foucault wrote:

The phenomenon unfolds calmly; it is inevitable, irresistible. . . . Watching it being born and grow, we realize that it is not in the experimenter's power to speed it up or slow it down. . . . Everyone who is in its presence . . . grows thoughtful and silent for a few seconds, and generally takes away a more pressing and intense feeling of our ceaseless mobility in space.[4]

Soon thereafter, the director of the Paris Observatory asked him to repeat the experiment in its *salle méridienne*, its central hall, located on the meridian. Foucault used the same bob but was able to lengthen the wire to 36 feet. This was preferable, because a pendulum with a longer wire swings for a longer time—it is less affected by friction in the air and in the mount where the wire attaches to the ceiling—and this magnifies the opportunity to see its apparent change of direction.

On February 3, 1851, exactly a month after starting his experiment, Foucault officially reported the results of his work to the French Academy of Sciences. The academy sent out dramatic invitations: *"Vous êtes invités à venir voir tourner la Terre, dans la salle méridienne de l'Observatoire de Paris"*—"You are invited to come watch the earth turn in the central hall of the Paris Observatory." At the gathering, Foucault told the crowd that most scientists studying pendulum behavior have focused on the time of their swing. Instead, his work had to do with the *plane* of their swing. Then, as his pendulum swung, he asked his audience to conduct a version of the thought experiment described above—to imagine building a pendulum "of the greatest simplicity" at the North Pole, setting its bob oscillating, and then leaving it "abandoned to the action of gravity." Because the earth "does not cease turning from west to east," the plane of oscillation appears to turn to the left, from the observer's perspective, as if the oscillation were attached to the heavens themselves.

Few scientific experiments met with such instantaneous fame as Foucault's pendulum. Although all educated Europeans in 1851 knew that the earth moved, all evidence for this fact—however incontrovertible—was based on inferences from astronomical observations. People without a telescope and the knowledge of how to use it had no way to see that motion for themselves. With Foucault's pendulum, the earth's rotation seems to become visible. A suitably educated person locked in a windowless room could prove that the room was rotating and, by careful measurement, even determine the latitude of the room.⁵ The pendulum, Foucault liked to put it, speaks "directly to the eyes."

Or does it? One of the fascinations of Foucault's pendulum is that it exhibits the ambiguities of perception. Foucault's remark is philosophically disingenuous: Nothing speaks directly to the eyes. The remark is Cartesian; Foucault imagines that his eyes are geometric eyes, and he has convinced himself that he can see what he imagines ideally and geometrically. If we can imagine the situation of the pendulum oscillating against the background of the solar system as a geometric model, he thinks, we can "see" the earth turn. But perception is more complicated than that. Even perceiving what is in motion and what is at rest depends on what we take as the foreground and what as the background or horizon. Foucault's pendulum seems to offer us the experience either of the pendulum turning in the gravitational field of the earth, or of the earth turning beneath our feet. This either-or seems to resemble French philosopher Maurice Merleau-Ponty's description of the familiar experience of being in a train stopped in a station beside another train on a nearby track—and when this other train begins to move we experience either that we are beginning to move or that the other train is beginning to move in the opposite direction. Which one, Merleau-Ponty writes, depends on where our perception is invested (in this train or the other), and what is its background or outer horizon.⁶ In order to see the plane of the pendulum's oscillation move, we need only do what perception habitually does, and take the object in question—the pendulum—as the foreground and the surrounding room as the background. Foucault's pendulum, like every instrument, shows what it does only within an appropriate environment. In order to "see" the earth moving, we would have to introduce a different and much bigger background in which the earth would show itself as moving, and the plane of oscillation of the pendulum would show itself as stationary. What if the pendulum were mounted not inside but outside? Could one see the earth turn on a starry night?

AS NEWS OF THE DEMONSTRATION filled Paris, Foucault was bombarded with correspondence from ordinary citizens, other scientists, and even interested government officials. Prince Louis-Napoléon Bonaparte—the President of the Republic, soon to become Napoléon III, Emperor of France—asked Foucault to set up a public demonstration in the Panthéon in Paris, a former church that had become the final resting place for many French national heroes. The Panthéon was, Foucault wrote, a marvelously appropriate location for the experiment, which now had a *splendeur magnifique*.[7] For the bigger the pendulum, the slower and more majestically it moved, and the more effectively it demonstrated the motion of everything around it. In an amazing display, Foucault attached a pendulum to the center of the Panthéon's huge dome. It had a 220-foot steel wire and a cannonball bob with a tiny, needle-like stylus attached underneath. Around the outer circumference of the circle where the pendulum was to travel, Foucault and his assistants built two semicircular banks lined with sand, which the stylus grazed at the extremity of each swing, marking the pendulum's position. In case the wire broke and the bob fell, Foucault protected the mosaic on the floor of the Panthéon directly underneath the dome by covering it with a layer of wood and several inches of densely packed sand. This was wise, for the first time the pendulum was installed the wire did in fact snap just below the dome, terrifying Foucault and his assistants as the 200-plus feet of wire whipped around the hall, convulsed with the pendulum's energy. When they reattached it, they installed a parachute up in the dome in case the wire broke again at the top.

On March 26, one of Foucault's assistants tied the bob to a wall

*Figure 7.1. Foucault's pendulum at the Panthéon, showing the banks
of sand to mark the pendulum's position at the end of each swing.*

with a cord and waited for it to come to rest. This time, the cord was
burnt through with a match rather than a candle (safety matches had
been invented that year). The pendulum moved ponderously, mag-
nificently, somberly, crossing twenty feet of floor with each swing,
making one back-and-forth oscillation every sixteen seconds. Its
thin wire, less than a millimeter and a half in diameter, was practically invisible against the grand setting, and the gleaming bob
looked like it was suspended in the void. When the bob reached the
banks at the end of its swing, the stylus cut a tiny furrow in the wet
sand, each about two millimeters to the left of the previous one. At
the latitude of Paris (about 49° N), the pendulum moved about one

degree every five minutes—a little over eleven degrees an hour, a rate that would carry it around a complete circle in about thirty-two hours, provided the pendulum did not first come to rest.

The Panthéon demonstration was not perfect; the furrow cut by the stylus slowly broadened into an extremely narrow figure-eight, evidently due to imperfections in the wire or the support. And the distance covered by the bob with each oscillation gradually shortened due to air resistance—though the time required for each swing remained the same (again, the principle of isochrony, discovered by Galileo, valid for all pendulums making small amplitude oscillations). Still, the pendulum continued to make the change of direction apparent for about five or six hours, in the course of which the direction shifted to the left in a clockwise direction (in the southern hemisphere, the direction would be counterclockwise) around the floor about 60 to 70 degrees. Enchanted, Louis-Napoléon rewarded Foucault by appointing him to a coveted position as a physicist in the Observatory, allowing him to move from his home basement laboratory.

The year 1851 was a year of marvels. The Crystal Palace exhibition in London opened, which marked a new era in display and visibility, and in the management of space and time. It was the first exhibition, for instance, at which audience tickets had times stamped on them, to manage traffic flow. Many historians, indeed, date the rise of modern mass society from this exhibition.

The year 1851 was also the Year of the Pendulum. Foucault pendulums proliferated all over the world: Oxford, Dublin, New York, Rio de Janeiro, Ceylon, Rome. Cathedrals—with their high ceilings and air of stability and authority—were perfect setup locations. In May 1851, one was set up in the cathedral of Nôtre Dame in Reims (forty-meter wire, 19.8 kilogram bob, more than 1 millimeter deviation each swing), one of the most beautiful Gothic cathedrals in France and the place where the kings of France were crowned. In June 1868, a Foucault pendulum was set up in the cathedral of Nôtre

Dame in Amiens, another Gothic masterpiece. And while the Crystal Palace exhibition was too long in the planning to exhibit a Foucault pendulum, one was featured at the Paris Exposition of 1855. For this event, Foucault invented an ingenious device that gave the bob a small electromagnetic boost on each swing to keep it from slowing down. That same year, his original pendulum was installed in the Musée des Arts et Métiers in Paris, an institution founded as a "depository for new and useful inventions," where it can still be seen.

But Foucault's pendulum was more than merely an interesting public demonstration. Like any scientific discovery, it reached back to the past and forward to the future. Researchers poring through the writings of earlier scientists uncovered evidence that others had noticed the direction of pendulums slowly drift to the left—including Viviani, the devoted disciple of Galileo, who had been the first to study pendulums seriously. Foucault, however, had been the first to connect this leftward drift with the rotation of the earth. Meanwhile, Foucault pointed out that the basic idea of his work had been anticipated by the late mathematician and physicist Siméon-Denis Baron Poisson (1781–1840). Poisson had calculated that cannonballs fired in the air should appear to veer slightly to the side as the earth rotated beneath them, though he had thought the deflection too slight to be observable. Poisson had also realized that the earth's rotation would affect pendulums—but had not grasped that the small effect on each movement of the bob would increase with each swing, allowing the motion, as Foucault put it, to accumulate the effects and allow them "to pass from the domain of theory into that of observation." Later, as the range of cannons increased, it became necessary for gunners to compensate for the effect described by Poisson. As physicist H. R. Crane noted,

> During the naval engagement near the Falkland Islands early in World War I, British gunners were surprised to see

that their salvos fell to the left of the German ships. They had followed the tables [for correcting their aim] prepared according to Poisson's formula, but had not remembered to change the signs for the corrections, to make them valid in the southern hemisphere.[8]

Foucault applied the same principle on which his pendulum had been based to invent the gyroscope, a word of his own coinage. A gyroscope consists of a spinning wheel mounted so that it can turn freely regardless of the direction of its support structure, and the spin axis of such a wheel always points in the same direction. Foucault predicted, correctly but decades prematurely, that it could and would be used as a directional device. The gyroscope principle, too, was found elsewhere in nature; scientists found, for instance, that common houseflies navigate with the aid of their own tiny twangers in the form of stiff stalks (rear vestigial wings) known as "halteres."[9]

Today, Foucault pendulums exist all over the world in science museums, universities, and other institutions. During the past half-century, many of these were made by the instrument shop of the California Academy of Sciences, which—in a specialized manufacturing service if there ever was one—has made nearly a hundred Foucault pendulums for institutions around the globe, including ones in Turkey, Pakistan, Kuwait, Scotland, Japan, and Israel. Often clients purchase the critical components, then embellish them with their own stylized interpretations.[10] The pendulum at the Boston Museum of Science moves back and forth over a brightly colored model of the Aztec Calendar Stone, with the bob crossing over the head of the Sun God Tonatiuh. The pendulum at the Lexington Public Library in Lexington, Kentucky, which was inaugurated with a cable-cutting ceremony at midnight on New Year's Eve 2000, has sensors in the floor to monitor the pendulum's motion instead of the usual pegs. The Montefiore Childrens' Hospital in New York had New York City artist Tom Otterniss design its bob and surrounding structure. The bob looks like an upside-down happy face, surmounted by a pointed conical hat that knocks over pegs. It swings over a silver-and-bronze relief map of the world centered on the Bronx, where the hospital is located. Meanwhile, little bronze sculpted characters, made of geometrical solids, are attached in various comical poses to the bob, wire, railing, and surrounding area. Nearly all visitors to the hospital stop to ask about it. Although Montefiore is but one of many institutions whose pendulum swings over a map centered on the building housing the display, the pendulum actually illustrates that every location on this rapidly spinning world is in motion—all are, to that extent, equal. Most appropriately, the United Nations headquarters in New York has a Foucault pendulum in the grand ceremonial staircase of its lobby, a 200-hundred-pound gold-plated sphere twelve inches in diameter that swings from the ceiling seventy-five feet above. The Smithsonian Institution, the United States national museum, used to have a Foucault's pendulum on display, but it was removed to make way for the restoration project of the Star Spangled Banner, the national symbol. The pendulum lies in a museum storage area.[11]

Like Young's experiment, Foucault's pendulum has to be executed with more care than it seems. In a public setting, a major problem is protecting a pendulum from visitors who seem to find irresistible the urge to reach out and touch it. And even though a pendulum is one of the simplest devices in science, a pendulum in the real world is affected by air currents, the internal structure of the wire, the way the wire is suspended, and how the bob is started; most of these things can easily throw off a pendulum or guide it into a figure-eight pattern. (A tip-off to a figure-eight pattern is that the knocked-over pins point inward.) At Stony Brook University, where I teach, a physicist demonstrating the Foucault principle to an introductory physics class once had a technician tie a bowling ball to the ceiling of the lecture hall, explained the principle to the class, and worked out how much of an angle it would travel during the forty-minute class time. At the end of the period, he measured the deviation and to his satisfaction found that it was exactly the calculated amount—but in the wrong direction! The erroneous amount was evidently due to some combination of a bad suspension system and the air currents in the drafty auditorium.

A Foucault's pendulum is quite different from other museum displays. Its sheer size is one factor: It cannot be enclosed in a booth or wall display but demands a huge open space like a nave or stairwell. It makes no sparks, hums, or noises but moves with a solemn majesty. And most important, it is not only noninteractive but seems to ignore us altogether, disclosing something radically counterintuitive to human experience. This—and its connection to vast physical forces—may be why people tend to remember their first Foucault's pendulum.

Mine put on the same performance every time I went to the Franklin Institute. But it never failed to enthrall me with its unsettling simplicity. It moved but stayed the same. It turned but told me that I was the one who was turning. I looked at it, and what it reflected back was my mobility and that of everything else around me—providing me with a clear and dramatic sense, whose true meaning I sensed I would never entirely fathom, of the deceptions and limits of my own perception and experience.

Interlude

SCIENCE AND THE SUBLIME

> *For beauty is nothing but the beginning of terror which*
> *we still are just able to endure and we are so awed*
> *because it serenely disdains to annihilate us.*
>
> —RILKE

FOUCAULT'S PENDULUM HAS what we might call a sublime beauty. Alongside the kind of beauty that presents us with clear visions and that integrates us with nature—making us feel more at home in the world—the sublime disconcerts because it confronts us with a terrible power. In the sublime, we experience our existence as puny and insignificant, and nature as incomprehensible and overwhelming: nature as an alien power.

The nature of the sublime was elaborated in the eighteenth century by such philosophers as Edmund Burke and Immanuel Kant, though they tended to contrast the beautiful and the sublime rather than to see the latter as a mode of the former. Burke wrote that the experience of the sublime is provoked by terror, "the ruling principle of the sublime." The terrible—and, for Burke, the terrible can be not only natural but also human-made, as in political terror—causes our usual strategies of coping with the world to crash down, and inspires "the strongest emotion which the mind is capable of feeling." But in a sublime experience (such as is made possible by artistic rendering) the terrible is held off at a safe distance so that we no longer feel in imminent danger. We can derive pleasure from such an experience, Burke continued, for we feel an ability to exist despite the terrible, to still be able to find a place for ourselves. This awareness not only gives us a small amount of a unique kind of pleasure, but also makes us more vibrant and alive.[1]

Writing shortly after Burke, Kant distinguished between two kinds of sublime. The first, or what he calls the "mathematically sublime," is associated with inconceivably vast magnitudes, such as the experience of the Pyramids or of St. Peter's Basilica in Rome, which make us feel like our imaginations are inadequate to grasp the whole. The other, the "dynamically sublime," is associated with overwhelmingly powerful forces: "Bold, overhanging, and, as it were, threatening rocks, thunderclouds piling up in the sky and moving about accompanied by lightning and thunderclaps, volcanoes with all their destructive power, hurricanes with all the devastation they leave behind, the boundless ocean heaved up, the high waterfall of a mighty river."[2] While for Kant our reason seeks to measure and control these, and all things, by producing categories adequate to them, in the sublime we experience the failure of this attempt—indeed, we grasp that our attempts to control such things will *never* succeed. Our sensibility has taken us off the deep end. This experience is unpleasant, but at the same time reveals the presence within us of a power (our own subjectivity) that these potentially physically annihilating things can never touch—a liberating revelation. The feeling of displeasure has its own pleasure—a perverse pleasure—since it makes us aware, through emotion rather than thought, of human freedom and the human transcendence of nature: things worthy of reverence.

Foucault's pendulum exhibits the sublime in science. It has little in common with Eratosthenes' experiment, which measures a length (the earth's circumference) already known to have some magnitude;

or with Galileo's inclined-plane experiments, which yielded a mathematical law; or with Newton's prism experiments, which uncovered a new aspect of nature. And all scientific experiments have a touch of sublimity, for they reveal that nature is infinitely richer than the concepts and procedures with which we approach it. But Foucault's pendulum highlights the sublime by the dramatic way in which it discloses the inadequacy of—or, rather, the mismatch between—human perception and the workings of nature.

The connection between Foucault's pendulum and the sublime, in its various manifestations from Burke and Kant to Umberto Eco, helps explain its celebrity at the time, and why it continues to be absorbing. It is not absorbing because it teaches us that the earth turns, or because it was a stepping-stone on the way to that indispensable navigational tool, the gyroscope. Rather, it is absorbing because it seems to summon unexpectedly deeper and even unfathomable truths about perception itself. Do we really "see" the pendulum move but "know" that it is in fact the earth moving? Or do we now, thanks to the pendulum and the explanatory sign by the staircase, really "see" the earth moving? In each of these two cases, science rules over perception, trumping it in the former case and correcting it in the latter.

Or, in experiencing Foucault's pendulum, is our perception being guided and retrained—not so much by the brute performance of the bob swinging back and forth in front of us, but by the explanations that are made for it, by the authority of the people making them, by the comprehensibility of the models that we are shown, by the way that these models integrate everything else that we know, and so forth? And if our perception can be as radically reeducated, as it is in this case, what other mysteries are out there? To what other things are we blind because of our particular perceptual education? What else might perception have in store? To be unsettled by such realizations is to experience the sublime.

Robert Millikan's oil-drop apparatus.

Eight

SEEING THE ELECTRON:

Millikan's Oil-Drop Experiment

—

WHEN AMERICAN PHYSICIST ROBERT MILLIKAN (1868–1953) delivered the customary address when he was awarded the Nobel Prize in 1923, he left his audience in no doubt that he had seen individual electrons. "He who has seen that experiment," Millikan said, referring to the experiment for which he had won the Nobel Prize, "has literally *seen* the electron."[1]

Millikan's stubborn insistence that his experiment allowed one literally to see subatomic particles was partly defensiveness—he was smarting from a dispute with another scientist who had challenged his work. But his claim to be able to see electrons was based on something different than Foucault's claim to be able to see the world rotate, because of the extraordinary environment provided by the equipment that Millikan had built in his laboratory.

WHEN MILLIKAN BEGAN his long series of experiments on the electron in 1907, he had spent more than ten years at the University of Chicago, had just received tenure, was married and had fathered three children, and was almost forty years old. Although he had written several well-respected textbooks, he had produced little important research, was hungry to make an original contribution to physics, and turned his attention to determining the electric charge carried by a single electron.

"Everyone was interested in the magnitude of the charge of the electron," he wrote in his autobiography, "for it is probably the most fundamental and invariable entity in the universe, though its value had never been measured up to this time with an accuracy even as good as a hundred per cent [i.e., the uncertainty is as big as the thing to be measured]."[2] Just as one of the main challenges in science during the eighteenth century was to measure the density of the earth—and then the gravitational constant—one of the main challenges to physics at the beginning of the twentieth was to measure the strength of the charge of the electron. And for the same reason—this information would tell much about the structure of the world.

In his Nobel address, Millikan introduced the subject of electricity by asking his audience to consider "a few simple and familiar experiments." If you rub a glass rod against a swatch of cat's fur and then touch a pith ball, the ball seems almost to acquire a "new and striking property" that makes it practically jump away from the rod. This, Millikan said, is an elementary phenomenon of electricity: Something called "electric charge" passes from the rod to the ball, and the ball and the rod repel each other in consequence. Benjamin Franklin had argued that that charge consisted of many tiny particles or atoms of electricity—that the phenomenon came in tiny grains or packages. By the end of the nineteenth century, scientists had proven to their satisfaction that Franklin was right: Charge was carried by minute bodies called "electrons," which somehow formed key parts of atoms. But they did not know if the electric charge of individual electrons came in packages of a specific size or could have any value whatsoever. That information was of vital interest to physicists interested in the structure of the atom, and to chemists interested in chemical bonds. But how could one find and measure the smallest grain of electricity?

Millikan knew that committing himself to measuring the charge of a single electron was chancy. He was giving up an established career as a textbook writer for a risky venture into research physics. He knew from his previous brushes with research "how much prospecting one could do in physics without striking a vein that had any gold in it." His goal—measuring the value of the electric charge of a single electron—was exceptionally difficult. Isolating and working with just one of these inconceivably small particles would be challenging enough in any circumstance. But it was not even clear at the time what was the best way to do the experiment. Millikan, so to speak, was not only trying to climb a high mountain, but trying to do so with only the roughest idea of which face would provide the easiest—or even a possible—ascent. Worse, the great scientific interest in the magnitude of the electron's charge meant that many people were trying to measure it. Millikan would be working in a crowded and competitive field, and the danger was great that others who were more experienced and better equipped could do it more quickly and accurately. He would need ingenuity and luck.

Millikan's biggest competitors were at the Cavendish Laboratory of Cambridge University. Its director, J. J. Thomson, had discovered the electron in 1897 (more precisely, Thomson discovered that all electrons had the same charge-to-mass ratio) and knew well the value of determining its exact charge; he was leading a group of talented students who were attacking the problem. They had tried many schemes, of which the most promising involved, surprisingly, creating a cloud of water droplets in the laboratory.

A few years previously, one of Thomson's associates had invented a device called a cloud chamber, which created clouds inside a chamber by causing supersaturated air (air that is full of water vapor) to condense on dust particles and free-floating particles containing electric charges called ions (negatively charged ions contain one or more electron charges). The fact that supersaturated air condensed around ions unexpectedly made the device useful for tracking the paths of rapidly moving charged particles such as those emitted by radioactive substances, for such particles left strings of ions in their wakes. In 1898, a year after his discovery of the electron, Thomson used this principle to make a rough estimate of the charge on the electron. He used a radioactive source to create negative ions (i.e., electrons) in the air inside a cloud chamber, then caused supersaturated air to condense around the ions—creating, in effect, a cloud of pith balls with charges on them—and measured the total charge of the cloud. He then estimated the total number of droplets in the cloud. This seemingly difficult task could be done, strangely enough, by measuring the rate at which the top surface of the cloud fell inside the cloud chamber. Thanks to an equation known as Stokes's Law, which describes the motion of tiny droplets through a fluid, Thomson could figure out the average size of the individual droplets making up the cloud by measuring the rate at which the cloud fell. (To do so, according to Stokes's Law, he needed to know the density of the droplets—easy, because they were made of water—and the viscosity of the medium they fell through—again, easy, because it was air.) Knowing the total volume of the water vapor in the cloud and the size of each individual drop then allowed Thomson to calculate the number of individual drops in the cloud. On the assumption that each water droplet in the cloud had condensed around a single electron, this information then enabled Thomson to divide the charge of the cloud by the number of droplets to approximate the charge of each electron.

Thomson's student Harold Wilson improved this method by

installing horizontal metal plates within the cloud chamber so that he could create an electric field inside the device. When he charged the plates, any charges in the region in between would be pulled down by the field. Using a stopwatch, Wilson measured and compared the rate at which clouds of droplets fell between a set of crosshairs, first under the influence of gravity alone and then under the influence of gravity plus the electric field, which pulled the cloud down a little faster. This was a significant improvement because it gave Wilson a way to make sure that the cloud layer he was measuring was composed of droplets with electrons in them, for the droplets with electrons would fall faster under the influence of the electric field than under the influence of gravity alone. It also allowed him to select the droplets with the smallest charge—for droplets that had condensed around more than one electron, having more charge, would fall at a faster rate. But Wilson's method, too, was only approximate, since the clouds evaporated quickly and successive clouds were often very different and hard to compare.

Millikan took up the matter in 1906 with a graduate student named Louis Begeman. They first tried Harold Wilson's method, but simply could not get it to work; the indefiniteness and instability of the cloud's upper surface made it next to impossible for them to measure anything with any accuracy. When Millikan reported this work at a meeting in Chicago, the eminent scientist Ernest Rutherford pointed out that one major difficulty was the swift rate of evaporation of the tiny water droplets. Millikan found that he had to drastically change his method to combat the evaporation problem, as well as a number of other difficulties.

Frustrated, Millikan decided to study the rate of evaporation so that he could compensate for it—another example of the "experimenter's vigilance" described in connection with Cavendish's experiment. Millikan decided to use a stronger electric field, and to reverse the current, so that it would pull the charged droplets up, holding the cloud steady while he examined its rate of evaporation. The first time he tried it, though, he got a shock that made him think, initially at least, that his goal was impossible and the experiment hopeless.

> When I had everything ready and . . . formed the cloud, I threw on the electrical field by turning a switch. *What I saw happen was the instantaneous and complete dissipation of the cloud—in other words, there was no "top surface" of the cloud left to set cross hairs upon* as Wilson had done and as I had expected to do.[3]

Virtually the entire cloud—evidently formed from droplets with more than one electron on them—had been swept away by the strong electrical field. This, Millikan wrote, "*seemed at first to spoil my experiment,* and with it all other experiments depending upon measurements of the rate of fall of an ionized cloud."

When Millikan repeated the attempt, the same thing happened. But suddenly he observed something that radically transformed his ideas—he saw that a handful of droplets still remained in view. "These were the droplets that chanced to have just that relation of charge to mass, or weight, which was needed in order that the downward force of gravity on the droplet might be balanced by the upward force due to the action of the field on the electric charge carried by the droplet. . . . Thus originated what I called '*the balanced drop*' *method of determining e* [the electron's charge]."[4]

Millikan had found a way to work with single pith balls, as it were, instead of clouds of them. By adjusting the strength of the electric fields in the chamber, he could make the tiny droplets move up and down in the chamber, and even stand still. After performing the experiment many times, he noticed that the charge needed to balance the drops was always an exact multiple of the smallest charge he observed on a droplet—providing the first unambiguous proof that electric charge did indeed come in grains.

Millikan then rebuilt the apparatus to study single drops instead of clouds. It consisted of a chamber in which charged water droplets fell through a tiny hole in a horizontal plate, whereupon they entered an area where their behavior could be observed with the aid of a microscope as they rose and fell between two cross-hairs.[5]

Millikan was phenomenally lucky in this experiment, and realized it. Only a narrow range of parameters made the experiment possible at all; if the droplets were much smaller, Brownian motion (the random motion of small particles suspended in a fluid due to collisions with the molecules of that fluid) would have made observing them impossible, while if the droplets were much larger, Millikan would not have been able to create the voltage required to hold them stationary. "Nature here was very kind," Millikan wrote later. "Scarcely any other combination of dimensions, field strengths, and materials, could have yielded the results obtained."

In the fall of 1909, Millikan sent in his first major paper on this "balanced-drop" method, which was published the following February. The paper is remarkable for the honesty of its presentation; in what science historian Gerald Holton describes as "a move rarely found in the scientific literature," Millikan included his personal judgment about the reliability and value of each of the thirty-eight observations of the droplets by ranking it. He triple-starred the two "best" observations, which had taken place "under what appeared to be perfect conditions"—meaning that he was able to watch the drop long enough to make sure it was stationary, was able to time its passage across the crosshairs, and observed no irregularities in its movement. He double-starred seven "very good" observations, single-starred ten "good" observations, and left unstarred thirteen "fair" observations. Also remarkable was that Millikan candidly said that he had discarded three "good" observations—whose inclusion would not have affected the final result—because something about their position or the field value made his measurement uncertain; three because of changes in the field value; and one simply because it was an outlier, with a value for its charge 30 percent lower than the others, which led Millikan to believe that there had been some experimental error. As Holton remarks, "Millikan was evidently saying he knew a good run when he saw one, and he was not going to overlook that knowledge even if it was not obvious how to quantify and share it on the record."[6] Such human judgments are always part of the scientific process, but experimenters rarely acknowledge them, and certainly not in print.

As if to confirm the principle that no good deed goes unpunished, Millikan would soon regret his honesty. That same year, a physicist at the University of Vienna named Felix Ehrenhaft (1879–1952) entered the debate. Using equipment similar to Millikan's but with tiny particles of metal instead of water droplets, Ehrenhaft claimed in 1910 that his results showed the existence of "subelectrons" with a range of charges smaller than those Millikan had found to be the smallest. Not only that, Ehrenhaft recalculated Millikan's own data, and, by using the observations that Millikan had thrown away as unreliable, made it seem as though the American experimenter's data actually supported Ehrenhaft's conclusions.

But by the time Ehrenhaft's paper appeared, Millikan realized how to improve his experiment vastly. In August 1909, shortly before submitting his first paper, Millikan had traveled to Winnipeg, Canada, to a meeting of the British Association for the Advancement of Science, whose president that year was J. J. Thomson himself. Although Millikan had not sought to be on the program, he brought his results, asked to speak, and attracted much attention when he presented them. Shortly after this meeting he decided to replace the water droplets with a heavier substance with a lower evaporation rate such as mercury or oil—a different kind of pith ball. In his autobiography, written twenty years later, Millikan described the breakthrough as a "eureka" moment that occurred on the return trip, realizing that it was foolish to try to combat the evaporation of water droplets when clock oils had been developed explicitly to resist evaporation.[7]

Like many such moments, however, how it really took place is as indistinct as the top surface of a cloud. In papers written at the time, Millikan credited his colleague J. Y. Lee with developing the atomizing method for producing tiny spherical drops for this experiment. And Millikan's graduate student Harvey Fletcher later claimed that he had originated the idea of using oil droplets. In all likelihood, nobody had a single breakthrough and there was no single origin of the "eureka" moment—the problem of curbing evaporation was uppermost in the minds of everyone involved with the experiment.

When Millikan arrived back in Chicago from Winnipeg, he hurried to his laboratory in Ryerson Hall, on the edge of the leafy quad in the center of campus. From the outside, one would hardly guess that the structure, a stunning neo-Gothic building with crenellated battlements, had been built as one of America's premier physics laboratories at the end of the nineteenth century. Even on the inside, its enormous oak beams and huge spiral staircase hardly spoke of a laboratory. It was sturdy, well insulated, and built of solid wood and heavy masonry—without iron, to avoid magnetic disturbances that might interfere with experiments involving tiny electric or magnetic fields. It had been built with input from the American physicist Albert Michelson, who insisted on certain specifications and building materials to facilitate his own experiments.

At the entrance to Ryerson, Millikan came across Michelson himself. Millikan told his eminent colleague that he had come up with a method that would allow him to determine the charge of the

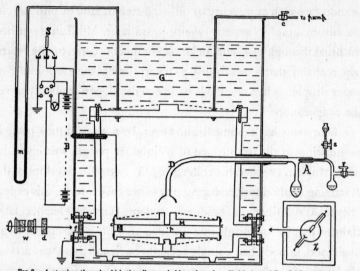

FIG. 8.—*A*, atomizer through which the oil spray is blown into the cylindrical vessel *D*. *G*, oil tank to keep the temperature constant. *M* and *N*, circular brass plates, electrical field produced by throwing on 10,000-volt battery *B*. Light from arc lamp *a* after heat rays are removed by passage through *w* and *d*, enters chamber through glass window *g* and illuminates droplet *p* between plates *M* and *N* through the pinhole in *M*. Additional ions are produced about *p* by X-rays from the bulb *X*.

*Figure 8.1.   Millikan's diagram, with original caption,
of his oil-drop experiment.*

electron to an accuracy of one tenth of one percent "or else I am no good." And he promptly went to the shop and ordered a new apparatus built suited to the balanced-drop method, but using oil. As before, he would create negative electric charges in a chamber full of droplets—oil, this time—choose one, and let it fall a few fractions of a second under the influence of gravity alone. From this he could calculate the radius of the drop. Then he would put a voltage on the plates and send the tiny droplet upward, then downward, then upward again. He observed the droplets through a small window, illuminated from the other side by a light. Measuring the droplets' times of rise and fall allowed him to calculate their charge.

From that moment on, Millikan devoted nearly all the time he could spare away from teaching duties to the experiment. His wife,

Greta, grew accustomed not only to his absence but also to making excuses for it when guests came for dinner. He was once puzzled when, after skipping a dinner for which he was supposed to be the host, he encountered one of the guests, who complimented him for being helpful around the home, which Millikan clearly was not. Later, it turned out that Greta had explained Millikan's nonpresence by saying he "had watched an ion for an hour and a half and had to finish the job," but the guests had understood her as saying that he "had washed and ironed for an hour and a half and had to finish the job."[8]

In September 1910, Millikan published a second major electron-charge paper—the first based on oil droplets—in *Science*. He had not yet seen Ehrenhaft's paper of a few months before, attacking Millikan's conclusions with his own data, and Millikan's second paper is written in the same vein as the earlier one. Although he did not rank the droplets, he forthrightly stated that he had not included some drops in his calculations of the electron's charge. In some cases, he said, this was due to their large experimental error— "when the velocities are exceedingly slow residual convexion currents [eddies in the air caused by heat] introduce errors, and when they are exceedingly fast the time determination becomes uncertain." In other cases, he omitted them because their values were "irregular," deviating from the norm by a large amount. Including these drops, though, would not significantly have affected the mean value of the electron's charge, only the degree of experimental error. Millikan wrote that "the method used is so simple, and the conclusions follow so inevitably from the experimental data, that even the man on the streets can scarcely fail to understand the method or to appreciate the results."[9]

Millikan continued upgrading his equipment—deploying a more accurate timepiece and better temperature control, for instance—and making more observations through 1911 and into

1912. In the spring of 1912, for instance, Millikan spent several weeks examining dozens of oil droplets, observing them through a microscope that he had installed in the wall of his chamber. On the afternoon of Friday, March 15, he spent about half an hour peering through the microscope at droplet number 41, using a stopwatch to time its rise and fall between tiny crosshairs. He could see this one very clearly, and the usual sources of disturbances, such as air currents, were absent. Despite the rather tedious work, he grew progressively more excited. When he finished recording the data in his laboratory notebook, he added in the lower left-hand corner the line cited in the introduction to this book: "Beauty. Publish this surely, beautiful!"[10]

By this time, Millikan had become aware of Ehrenhaft's paper—and several even more vehement successors—which charged that Millikan's paper was false, and that his own data demonstrated the existence of subelectrons. In 1913, Millikan published a comprehensive paper based on work with his improved equipment. Clearly a little stung by Ehrenhaft's charge, he reported defensively that the data came from a set of observations with fifty-eight droplets, which, he wrote pointedly, was not "a selected group of drops but represents all of the drops experimented upon during 60 consecutive days."[11] This work, he wrote, established the value for the charge of the electron ($4.774 \pm 0.009 \times 10^{-10}$ electrostatic units, or esu) to 0.5 percent.

The scientific community accepted Millikan's results, based not only on this paper but also on other evidence supporting the atomistic character of electricity, and he was awarded the Nobel Prize in 1923 partly for this work. For a few years Ehrenhaft continued to press his case for subelectrons, but eventually he gave up. Later in his career Ehrenhaft became obsessed with another cause—magnetic monopoles, which can be thought of as magnets that only have one end. (They may exist, but nobody has ever seen one.) Occasionally Ehrenhaft showed up at scientific meetings clutching what he said was evidence for them. A poignant moment occurred in 1946 at an annual meeting of the American Physical Society in New York City. The young theorist Abraham Pais gave a presentation that was interrupted by Ehrenhaft, then pushing seventy and still championing the monopole cause. He approached the podium demanding to be heard, and was politely escorted out of the room.

A young physicist named Herbert Goldstein was sitting next to his mentor, Arnold Siegert. "Pais's theory is far crazier than Ehrenhaft's," Goldstein said to Siegert. "Why do we call Pais a physicist and Ehrenhaft a nut?"

Siegert thought a moment. "Because," he said, "Ehrenhaft *believes* his theory."[12]

The strength of Ehrenhaft's conviction, Siegert meant, had interfered with the normally playful attitude that scientists require, an ability to risk and improvise. (Conviction, Nietzsche said, is a greater enemy of truth than lies.)

WAS EHRENHAFT RIGHT that Millikan was fudging his data? Holton's examination of Millikan's lab notebooks for the work on which his 1913 paper was based revealed that Millikan had in fact studied 140 drops—not 58, as he had claimed. Millikan's statement that "this is not a selected group of drops but represents all of the drops experimented upon during 60 consecutive days" was therefore false. Although bound to raise eyebrows, this revelation did not unduly trouble Holton himself. Holton suggests two partial explanations. One was the controversy with Ehrenhaft; Millikan, convinced he was right, did not want to give Ehrenhaft any more ammunition, which in his eyes would only confuse the issue. The second reason Millikan omitted mention of the missing drops is clear from the sources of experimental error that Holton found recorded in the notebooks themselves: "The battery voltages have dropped; the manometer is air-locked; convection often interferes; the distance must be kept more constant; stopwatch errors occur; the atomizer is out of order." Millikan, in short, did not believe that the "missing" eighty-two drops were data at all. Millikan's notebooks note the difference between drops observed under perfect conditions, which he often describes as "beautiful," and drops where the observations were affected by various degrees of experimental error. Here is a sample that Holton extracted from Millikan's notebooks for the last week of the run:

Beauty. Tem & cond's perfect, no convection. Publish [April 8, 1912]. Publish Beauty [April 10, 1912]. Beauty Publish [crossed out and replaced by] Brownian came in [April 10, 1912]. Perfect Publish [April 11, 1912]. Among the very best [April 12, 1912]. Best one yet for all purposes [April 13, 1912]. Beauty to show agreement between the two methods of getting $v_1 + v_2$ Publish surely [April 15, 1912]. Publish. Fine for showing two methods of getting v . . . No. Something wrong with the therm.[13]

Millikan therefore picked and chose the droplets that he published, and to avoid fueling further criticisms from Ehrenhaft did not report that he was omitting them, viewing them as irrelevant to the real issue of the electron's charge. To use Holton's image, Millikan exercised judgment in discriminating what he let in the scientific "window"—what he accepted as data. By contrast, Holton wrote, Ehrenhaft and assistants "appear to have used all their assiduously collected readings, good, bad, and indifferent." They let everything in the window, and treated it with equal value.

Ever since Holton's article, historians, journalists, and scientists have debated the validity and ethics of Millikan's procedure. Most of the time, they package the Millikan story to exhibit a lesson, cleaning up and trimming the account to make the point clear—creating, in effect, historical demonstrations. To some extent this process occurs in all writing about historical events, but in Millikan's case it is particularly interesting. Science historian Ullica Segerstråle has dryly described what has happened to the story of Robert Millikan's Nobel Prize–winning experiment as a case of "canned pedagogy."[14] What is remarkable in this case is the polar opposition of the cans that the episode is mashed into: Millikan the brilliant scientist on the one hand, Millikan the exemplar of shameless fraud on the other.

For obvious reasons, some journalists and science writers, making a quick read of Holton's article, have focused on Millikan's omission of drops—and especially on his false claim in the 1913 paper that he reported all his observations. In their view, the Nobel laureate was guilty of scientific misconduct and even fraud.[15] In their 1983 book, *Betrayers of the Truth: Fraud and Deceit in the Halls of Science*, *New York Times* journalists William Broad and Nicholas Wade thundered that "Millikan extensively misrepresented his work in order to make his experimental results seem more convincing than was in fact the case."[16] And medical doctor Alexander Kohn includes Millikan among his "false prophets" of science in his book of the same title, though Kohn seems more exercised by Millikan's alleged neglect of the contributions of his graduate student Fletcher than by the omission of data.

On the other side, several historians of science, focusing on the fact that Millikan used what now looks to be sound judgment about the reliability of his data, have praised him as an example of the good scientist. These scholars point out that scientific thinking is often not a matter of numbers but of judgment, and point to numerous historical cases of scientists correctly interpreting experiments where strict reliance on numbers alone would have led them astray. When it comes to data, all numbers are not created equal. In 1984, science historian Allan Franklin painstakingly analyzed each of the drops Millikan omitted in his 1913 paper and showed that almost all were indeed omitted for reasons of experimental error—and, perhaps even more important, that even if Millikan *had* included them, the final result would not have greatly changed.[17]

These stories tend to get recycled by those who are more interested in their favored lesson than in historical accuracy or in the scientific process. Each version omits complexity. The Millikan-as-bad-scientist version leaves out the reasons why not all data are good and therefore why it is often wise to discard some, while the Millikan-as-good-scientist version leaves out the pressures to get a result first, and therefore to compromise reporting of data. As Segerstråle pointed out, the clash to a large extent stems from the application of two very different, and largely incompatible, ethical perspectives to the scientific process. In one, the Kantian (or "deontological") perspective, ethical behavior consists of the intention to apply to oneself the same rules that everyone else does to themselves—and Millikan was bad because he didn't follow the rules about reporting data. In the other, utilitarian perspective, what science is about, from beginning to end, is simply to get the right result—and that is what Millikan did. In fact, as Segerstråle points out, science is so competitive that those who do not race ahead to get the quick-and-dirty right result tend to drop out.

The argument about Millikan's conduct has made it difficult to recover the beauty of his experiment—but it is well worth the effort to do so. For that, we have to ask ourselves what Millikan actually *saw*. He would peer through a microscope into a chamber that he had designed himself. That chamber was a little stage for a peculiar kind of action carried out by a peculiar kind of actor. The actors that appeared one at a time on this little stage were tiny oil droplets a few microns across. This is so tiny—their diameter was about the wavelength of visible light—that light actually curled around them and you could see its diffraction. They did not look solid in the crosshairs but rather like smeared discs surrounded by diffraction rings—which is why Millikan could not measure their size optically and had to resort to Stokes's Law to determine their size. Each drop, when illuminated by an arc lamp, looked to Millikan like a twinkling star on a dark sky. The droplets were extremely sensitive to the environment, and responded to any currents in the air, to collisions with air molecules, and to the electric fields that Millikan would adjust to make them move. He saw the drops go up and down in response to the changing electric field. He saw them drift in other directions thanks to the air currents. He saw them jiggle back and forth due to Brownian motion. He would watch a droplet move in the electric field, and then suddenly see it jump when it encountered another ion in the air. "One single electron jumped upon the drop. Indeed, we could actually see the exact instant at which it jumped on or off."[18] When an oil droplet was "moving upward with the smallest speed that it could take on, I could be certain that just one isolated electron was sitting on its back." He knew how to make the drops go up or down or stay absolutely still. He became familiar enough with them to recognize everything that was happening— and that what was happening was showing him something new about the world. There is a sensual pleasure to seeing objects behave in complex situations according to laws one knows intimately—like that of watching a basketball sail through the air, bounce on the lip of the hoop, against the backboard, and then back down through the hoop. Only what Millikan was seeing here was an action that showed him something fundamental—the fundamental electric charge. It was the kind of beauty Schiller was speaking about, something that "conducts us into the world of ideas without however taking us from the world of sense."

On a whim one afternoon in Chicago, I decided to try to find the spot where Millikan had performed his famous, Nobel Prize–winning series of experiments to measure the charge of the electron—a defining moment of our electronic age. I went to the University of Chicago and found my way to Ryerson Hall, but was unable to locate a commemorative display. Nor could I find anyone wandering around the halls who could tell me the room where the experiment had taken place; some even asked who Robert Millikan was. A secretary suggested that I call public relations, but they, too, were at a loss. I found no trace of Millikan or his experiment in the building, which was now home to the Department of Computer Science. Lab demonstrations and historical packagings and repackagings will persist—but Millikan's actual experiment, like most science experiments, has faded into the woodwork.

Interlude

PERCEPTION IN SCIENCE

SCIENTISTS OFTEN SAY that they "see" the objects they work with, however tiny or abstract these may be. Biologist Barbara McClintock once remarked, in connection with her research on chromosomes, "I found that the more I worked with them, the bigger and bigger [the chromosomes] got, and when I was really working with them I wasn't outside, I was down there. I was part of the system. I was right down there with them, and everything got big. I even was able to see the internal parts of the chromosomes."[1] Astronomers often speak of "seeing" a planet circling a pulsar when, for instance, they pick up fluctuations in the radio signals emitted by the pulsar caused by gravitational effects created by the orbiting body. And soon as a sodium cloud was discovered issuing from a volcano on Jupiter's moon Io a few years ago, an astronomer was quoted as saying that it was "the largest permanently visible feature in the solar system."[2]

Such remarks may sound like little more than loose talk, on the same level as, "I see that it's going to rain," and not really to involve "seeing" at all. Aren't true scientific entities—from electrons to black holes—imperceptible, accessible only through some form of instrumental mediation?

Whether scientists really perceive what they study is important to the question of science and beauty. For most descriptions of beauty emphasize that it involves sensible perception; apprehending something immediately and intuitively. If scientists work only with abstractions, inferences, and equations, such sensible perception would be all but impossible.

Perception in science is a fascinating and complex subject, but it is not different in kind from ordinary perception.[3] In ordinary perception, after all, we do not see only shapes or swaths of colors—green pears and yellow pencils—but far more complex phenomena, including things like examples of courage and intelligence, self-deception and addiction, gambling and ambition. How is this possible? A basic phenomenological principle, as I mentioned in the previous chapter, is that what we perceive is not automatic or preordained but depends on what we take as the foreground and what as the background or horizon. What we perceive, one might say, is "read" like a sign system against that background. In ordinary perception, the background is generally given—but in science we are able to change the background thanks to reliable instruments and technologies, and thus to perceive new things. This can be as simple as seeing which way the wind is blowing, or what the temperature is, by looking at a wind vane or a thermometer. But it can also be more complex, as in the case of seeing electrons in cloud chamber tracks or various anatomical features in X rays—which nonscientists can and have been taught to do. In the early days of high-energy physics, before computers took over, physicists hired housewives and even liberal-arts graduate students to identify muons and pions and other kinds of particle tracks. Not only does human perception always take place against a background, but it is also subject to education.

Whenever we perceive an object, we grasp a certain regularity or invariance in its appearances (or profiles, as the philosophers say). For me to perceive an object as a desk and not as an illusion, cardboard prop, or sculpture is to know that if I walk around it I will see another side not now visible to me, and that I then no longer will see *this* side—and that throughout all those changes I would still be seeing it as the "same" object. This implicit horizon of appearances that "comes with" my seeing something as an object is not speculation or guesswork on my part; it is what it means to see an object. If I suddenly think I see the president of the United States standing in the middle of the sidewalk in front of me, I might step to the side to obtain a profile from a new angle that shows me that the object is instead a photographer's prop, in which case I no longer see it as a person but as a cardboard display.

To perceive either an ordinary or a scientific object is to grasp a particular profile of an object together with such a horizon of expected profiles. This is true even when we see something as ordinary as an apple. In every successive experience—picking it up, turning it over, biting it—we gain increasing fulfillment of its horizon of profiles. We can be surprised—the apple can turn out to be made of wood or glass, say. But we perceive this in an experience that reshapes but does not eliminate the horizon of profiles.

While invariances in ordinary perception are intuitions of physical regularities, those of scientific objects are usually described through theories. To see a chromosome, planet, sodium cloud, or other scientific object is to understand that object as obeying certain regularities or invariances—defined by whatever theory describes that object. Whether or not we continue to see these phenomena as such will depend on how their profiles fulfill the expectations raised by these invariances.[4]

"Wonder" is the name we give for the desire to explore the given and promised profiles of a phenomenon for its own sake—to engage in an adventure of fulfillment—and it is observable not only in humans but primates and other creatures. Wonder is thus "certainly not a mere social construction," writes philosopher Maxine Sheets-Johnstone, but part of our evolutionary lineage. The scientific temperament pursues this adventure through experiment, which produces new and often unexpected profiles of phenomena.

In the case of most objects—cups, chairs, even people—we know pretty much what to expect in the horizon of profiles. Occasionally, though, we have not only the awareness but even the expectation of the possibility of surprise. These things we call "mysterious." The scientific temperament involves openness to the possibility of being surprised. This is surely behind Einstein's remark that "the most beautiful thing we can experience is the mysterious. It is the source of all true art and science."[5]

In laboratories, we can create special background environments with reliable instruments and technologies—from thermometers, X rays, and NMRs to complex particle detectors—in which new things show themselves. Millikan's apparatus was one example. Inside was a world of its own, and Millikan became thoroughly familiar with it. He knew its laws and disturbances. He recognized typical behaviors and situations in that world, and he could also recognize atypical behaviors and situations, when he knew he did not understand everything that was happening. For this reason, it is appropriate to say that he could *see* things in that world.

This kind of familiarity is what scientists from McClintock to Millikan have with the subjects of their research—their ability to grasp the world in which they work well enough to see objects in it is indeed a condition for being able to find beauty in it.

Rutherford's first rough note on the nuclear theory of atomic structure, written probably in the winter of 1910–11.

9

DAWNING BEAUTY:
Rutherford's Discovery of
the Atomic Nucleus

—

IN THE FIRST DECADE OF THE TWENTIETH CENTURY, an ingenious experiment allowed the British physicist Ernest Rutherford (1871–1937) to discover the internal structure of the atom. To scientists' surprise, he learned that atoms consist of a central, positively charged core or "nucleus" that contains nearly all the mass of the atom, surrounded by a cloud of negatively charged electrons. Until then, the ultimate structure of matter was one of those mysteries—like the beginning (and end) of the universe, the origin of life, and the existence of life on other planets—that were interesting subjects for speculation but impossible to investigate. How, scientists wondered, could they study the internal structure of atoms if the only tools they had were made of atoms themselves? It would be like trying to find out what was inside a rubber ball—with another rubber ball. Rutherford's achievement marked the birth of modern particle physics.

The route to Rutherford's discovery was anything but straightforward. He did not start out to find the structure of the atom. It took him time to realize that he had a tool for carrying out such an experiment, to figure out the right way to use that tool, and to understand what the experiment was telling him. And it took others time to become convinced.[1]

RUTHERFORD WAS A LARGE, confident man with a ruddy face, a walrus mustache, a loud laugh, and a booming voice, who kept pushing his assistants and collaborators to keep things simple. To explain his successes, Rutherford liked to say, "I am always a believer in simplicity, being a simple man myself."[2] This was not all bluster. He understood the power of simple equipment to trick nature into revealing its most profound secrets.

Indeed, for their simplicity, profundity, and definitiveness, Rutherford's experiments are some of the most beautiful in science. His colleague and sometime competitor J. G. Crowther wrote of being astonished that Rutherford's simple ideas embodied in his experiments could still be effective in the twentieth century: "One might have expected that after three centuries of intensive development of physics, ideas would necessarily have evolved a complicated subtlety and all simple ones would have been used up and exhausted."[3] And according to another colleague, A. S. Russell, "On a backward view one saw the beauty of the method of investigation as well as the ease with which the truth was arrived at. The minimum of fuss went with the minimum chance of error. With one movement from afar Rutherford, so to speak, threaded the needle first time."[4]

Rutherford seems to have had little appreciation for art. As for his musical tastes, when he burst into song his choice "was usually an off-key version of 'Onward Christian Soldiers' rendered with great gusto."[5] But his approach to wresting the underlying structure of the world into the light had all the hallmarks—fierce energy, deep respect for the material, strong physical imagination—of the good artist. Indeed, Rutherford once argued that "the process of discovery may be regarded as a form of art."[6]

But as in art, so in science: The creative process often is convoluted, backtracking is common, and artists often realize what they are looking for only at the end. A classic illustration is Rutherford's masterpiece, the discovery of the atomic nucleus.

Rutherford was born in New Zealand, and as a youth tinkered with cameras, clocks, and small models of the waterwheels at his father's mill. In 1895, he was awarded a special fellowship that took him to England and to the Cavendish Laboratory, which science historian J. L. Heilbron has called "the nursery of nuclear physics."[7] He arrived at the beginning of an exciting and demanding time in physics: German physicist Wilhelm Röntgen discovered X rays in 1895, French physicist Henri Becquerel discovered radioactivity in uranium in 1896, and British physicist J. J. Thomson—the director of the Cavendish Laboratory—discovered the electron in 1897.

Rutherford easily distinguished himself in this intense atmosphere, and in 1898 left the "nursery" to accept a professorship at McGill University in Montreal, where he would remain until 1907. Just before leaving, while investigating radioactivity, he made the unexpected and crucial discovery that uranium emitted two different kinds of radiation. Typically, he devised a simple and utterly convincing experiment to show this: He covered uranium with layers of aluminum foil and measured the amount of radiation that came through. One or two layers cut down the amount, but at three layers the level of radioactivity dropped noticeably. Strangely, the remaining radiation was not significantly blocked by a fourth or even fifth layer. It continued to come through until Rutherford covered the uranium with many layers of aluminum. To Rutherford, this showed that uranium emitted two types of radiation, one significantly more powerful than the other. He called the less penetrating kind "alpha rays" and the more penetrating kind "beta rays," after the first two letters of the Greek alphabet.

As it turned out, alpha rays—what they were, how they behaved, and to what use they could be put—would become the focal point of Rutherford's career. Rutherford's students liked to say that the alpha particle was a little creature that their mentor had brought into being by accident and then made all his own. They would work wonders together, Rutherford and his little creature. It would become his tool for unlocking the interior of the atom, though he would discover that, too, by accident.

Rutherford quickly learned that neither alpha nor beta rays were actually rays in the sense of, say, X rays. Instead they were little bits of electrically charged matter that uranium atoms spat out of themselves for then-unknown reasons. Beta rays were negatively charged, and soon shown to be electrons, but the nature of alpha rays, which were positively charged, was initially a puzzle. Rutherford solved it. He had noticed that their mass was similar to that of helium atoms—but were they helium atoms? He devised another ingeniously simple demonstration to prove it. He had a glassblower prepare a glass tube with walls thin enough to allow alpha rays to pass through, yet strong enough not to crumple under atmospheric pressure. He filled this tube with radon, a gaseous element known to emit alpha rays, and surrounded it with another airtight glass tube, leaving an empty space in between. He pumped out all the air from this space, leaving a vacuum; the only thing that could enter it were alpha rays passing through the walls of the inner tube. Rutherford found that a gas slowly collected in that space at a rate proportional to the rate at which the alpha particles passed through the inner wall. He then tested the gas and showed that it was helium. Alpha rays—or alpha particles, as they were increasingly called—were helium atoms. "This experiment," wrote

Rutherford's student Mark Oliphant, "created great interest on account of its simple directness and beauty."[8]

Mysteries remained. How did the positively charged alpha particles turn into helium, which ordinarily was electrically neutral? What were helium atoms doing inside uranium atoms, anyway? Were they pieces that chipped off an atomic block, or something else? How were they related to the rest of an atom's nucleus? Rutherford's path to the answer to this puzzle was indirect. It began with a friendly quarrel with Becquerel, some of whose experiments with alpha particles were at odds with Rutherford's. After noticing their conflicting results, the two men each looked into the matter more closely, and Rutherford was proven right. But the dispute sparked his curiosity: Why *was* it so maddeningly difficult to measure the properties of alpha particles? How had Becquerel, whom he knew to be careful, been misled? The reason was the alpha particles' habit of ricocheting off air molecules.

Rutherford was familiar with this behavior, which he demonstrated in his usual simple and direct style: He first fired a beam of alpha particles onto a photographic plate in a vacuum, obtaining a crisp, bright spot at the point of impact. Then he shot the same beam at the same plate, not in a vacuum this time but through air. This time the spot spread out and was blurred. The smearing out of the spot, Rutherford wrote in 1906, was due to a "scattering of the rays" as they bounced off molecules in the air. Though Rutherford did not know it yet, the discovery of the role of scattering was a key step en route to the discovery of the nucleus.

Two years later, Rutherford was awarded the Nobel Prize, strangely enough in chemistry rather than physics, for his "investigations into the disintegration of elements and the chemistry of radioactive substances." During the ceremony, he quipped that he had seen many transformations in his work, but the quickest ever was his own from a physicist to a chemist. By this time, he had moved back to England, to the University of Manchester. And as he grew ever more interested in measuring precisely the various properties of alpha particles, he was growing ever more frustrated by the scattering. It was seriously affecting, for instance, his ongoing attempts to measure the charge of alpha particles by shooting them one by one into a detector. His colleagues were also troubled by scattering, and his colleague W. H. Bragg sent him some drawings of tracks "with elbows on them" that alpha particles had left in cloud chambers. "The scattering is the devil," Rutherford complained in a letter to another colleague.

Exasperated, Rutherford asked his new assistant, Hans Geiger, to measure the scattering. (Geiger later invented the famous Geiger counter, which detected radioactivity electronically in laboratories and in countless postwar thrillers.) It was yet another example of experimenter's vigilance—the same instinct that had led Cavendish to measure the strength of the magnetic fields in his torsion-bar apparatus and Millikan to study the evaporation of water droplets. If you have a disturbing force in your experiment, first measure it directly, then compensate for it. As it happened, Rutherford's request to Geiger was another key step on the path to discovering the atomic nucleus. This, too, Rutherford did not realize at first. To him, it seemed simply as though he were being forced into trying to understand and quantify a disturbance that was muddying the precision of his measurements of alpha particles' charge and mass.

Measuring alpha particles was a chore. Rutherford and Geiger had learned that when alpha particles strike certain kinds of chemicals, such as phosphorescent zinc sulphide, they create tiny momentary flashes, known as "scintillations," that can be seen with a microscope. It was the first time that single atoms (alpha particles being counted as helium atoms) had been detected visually. By watching screens painted with such chemicals, scientists could establish where alpha particles hit the screen, providing information about their trajectories. But to observe the faint, ephemeral scintillations, Geiger had to sit in the dark for at least fifteen minutes to adjust his eyes enough to see the flashes. It was time consuming and tedious.

The equipment Geiger used to measure the scattering was simple by today's standards. A little bead of radium—an intensely radioactive element that spat out alpha particles in a near-continuous stream—sat in a small metal canister. The canister was rigged with slits to beam a narrow line of alpha particles down a glass tube about four feet long. All the air was pumped out of this firing tube so that the alpha particles would not be scattered by air molecules. Connected to this firing tube was another, similar tube, also without air, through which the alpha particles would pass before encountering a zinc-sulfide screen. By peering through a microscope trained on the screen, Geiger could observe the flashes and measure their positions. Almost invariably, these flashes would all occur at the same spot. Then Geiger placed thin pieces of metal foil between the first and second glass tubes. Now the flashes were not all in the same place, and some seemed to dance over the screen.

Geiger explained what was happening in a presentation to the Royal Society in June 1908. Most of the alpha particles sailed right through the foils, he said, but now and then one was scattered by them. Like a cue ball brushing a stationary ball on a pool table, the alpha particle had been knocked to one side. Moreover, the thicker the foil, the greater the number of alpha particles scattered and the greater the angle at which they careened off. Evidently, these alpha particles had collided with several atoms when they passed through the thick foils. In addition, foils made from heavier elements, such as gold, scattered alpha particles more than foils made from lighter elements, such as aluminum.

It was difficult for Rutherford and his co-workers to picture what was happening in the scattering. Alpha particles, they knew, shot out of radium at tremendous velocity—on the order of 10,000 miles per second. It was hard to imagine how the atoms in the thin foil could deflect such hugely energetic entities. Rutherford and his co-workers, in fact, did not yet have the modern picture of alpha particles as billiard balls or bullets; all they knew was that alpha particles were essentially atoms—helium atoms—but they knew nothing about the structure of these atoms. The discovery that some atoms, at least, emitted positively charged alpha particles and negatively charged beta particles had inspired a few scientists to begin thinking about the internal structure of atoms (including alpha particles/helium atoms). Atoms surely contained electrons. And since ordinary atoms are electrically neutral, they contained a positive charge as well. But how, and in what form? In 1904, J. J. Thomson had proposed that an atom consisted of electrons held together by a positive jelly—like plums in a pudding, it was said, leading this picture to become known as the plum-pudding model. That same year, a Japanese scientist proposed a planetary model, the atom consisting of a central core and surrounding satellites. But because these remained guesses, it was difficult to picture what was happening when an alpha particle/helium atom caromed off another kind of atom.

Still trying to understand the scattering, Geiger continued on, working now with an undergraduate assistant, a New Zealander named Ernest Marsden. Through the fall of 1908 and spring of 1909, Geiger and Marsden improved the apparatus, inserted washerlike pieces to cut down on particles scattered from the walls of the tube, and used a more intense beam, but still could not obtain consistent measurements. The trouble seemed to be that the alpha particles were deflected not only by the foil but by residual air in the tubes as well as by various parts of the tube and the rest of the experimental apparatus. With so much bouncing around, it was difficult to tell what was being scattered off what.

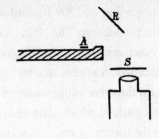

Figure 9.1.
Sketch of Geiger
and Marsden's
experimental
setup to measure
wide-angle
scattering.

One day in the early spring of 1909, Rutherford, who had been following Geiger and Marsden's work and its growing tribulations, entered their laboratory, and, in Marsden's recollection, said, "See if you can get some effect of alpha particles directly reflected from a metal surface." Rutherford wanted them to rearrange the experiment to see if the alpha particles would bounce directly off the foil, like a tennis ball bouncing off a wall, instead of being scattered as they passed through it. Again Geiger and Marsden deployed a simple experimental setup. They moved the screen off to the side and shielded it with a lead plate to block off all alpha particles from reaching the screen except those that would have rebounded off the metal foil (Figure 9.1). They had to increase the intensity of their source still further to maximize the number of particles traveling at large angles. Almost immediately they learned that some particles were indeed shooting off to the side. After several weeks of experimentation, using different kinds of metal foils and foils of different thicknesses, they found that about one in eight thousand alpha particles was reflected by an angle of greater than ninety degrees. "At first," Geiger wrote years later, "we could not understand this [wide-angle scattering] at all."[9]

By this time, Rutherford had realized, to his annoyance, that if alpha particles were scattered by one or more chance encounters with atoms, to understand this process he would have to learn much more about the mathematics of probability than he knew. As a result, at the beginning of 1909 Rutherford enrolled in an introductory course in probability. The Nobel laureate diligently took notes and did the exercises, and eventually managed to work out a theory of what he called "multiple scattering," which applied to cases where the particles were being scattered by chance encounters with several atoms, each of which scattered the alpha particle by a small amount. But the theory of multiple scattering did not seem to fit the wide-angle scattering that Geiger and Marsden were now finding.

At a lecture given at the end of his life, Rutherford spoke about the time that Geiger and Marsden first set up their experiment:

> Then I remember two or three days later Geiger coming to me in great excitement and saying, "We have been able to get some of the alpha particles coming backwards. . . ." It was quite the most incredible event that has ever happened to me in my life. It was almost as incredible as if you fired a 15-inch shell at a piece of tissue paper and it came back and hit you.[10]

Rutherford's incredulity is a case of experimenter's hindsight. In physical terms, it *was* that incredible—a heavy alpha particle, fired at about ten thousand miles a second, bouncing off a wispy slip of foil! But even Rutherford's prodigious physical imagination did not quickly grasp just how incredible it was.

Initially he held to his thought that the wide-angle scatters could be accounted for by multiple scattering—that is, that the alpha particles must have collided with an extremely large number of atoms—and that this somehow ended up kicking them backward. But over the course of the next year, as he worked on probability theory and digested the results of this experiment as well as some additional developments, his thoughts began to change. One of these developments was his growing conviction that an alpha particle was not a blob or pudding, but could be treated like a point. This was a huge step, for among other things it vastly simplified the mathematics of scattering theory. It also helped him realize how valuable a tool alpha-particle scattering was. If you knew enough about the scattering, and learned how it was affected by various parameters, such as charge and mass distribution, you could then reverse the process and find out, by the way alpha particles scattered, information about the scattering medium. Scattering was thus not just an unpleasant effect that experimenters had to live with, but an interesting phenomenon that could tell you something about other things.

In particular, it was dawning on Rutherford that alpha-particle scattering could tell him something about the structure of the atom itself. According to Geiger, Rutherford had one key insight shortly before Christmas 1910.

> One day Rutherford, obviously in the best of spirits, came into my room and told me that he now knew what the atom looked like and how to explain the large deflections of alpha particles. On the very same day I began an experiment to test the relations expected by Rutherford between the number of scattered particles and the angle of scattering.[11]

And one of Rutherford's Sunday dinner guests, Charles G. Darwin, recalled Rutherford exuberantly saying at that time that "it is really very fine to see the things one has seen in imagination visibly demonstrated."[12]

The simplifications of scattering theory had helped Rutherford realize that the alpha particles could not be explained by multiple scattering—they were not turned around by multiple collisions, but only from a single collision. This in turn could only happen if almost all the mass of the atom were concentrated in a single charged node in its center.

What Rutherford evidently saw in his imagination was that the atom consisted of a massive charged core, surrounded by mostly empty space—emptier still than the solar system. If an atom were blown up to the size of a football stadium, the nucleus would be the size of a fly at its center and electrons even tinier specks distributed throughout the rest of the enclosure. Virtually the entire mass of the stadium, however, is contained in that tiny core. But Rutherford was still unclear whether it was positively or negatively charged. In March 1911, he wrote to a colleague: "Geiger is working out the question of large scattering and as far as he has gone results look very promising for the theory. The laws of large scattering are completely distinct from the small scattering. . . . I am beginning to think that the central core is negatively charged."[13] The positively charged alpha particles, he evidently thought, were swinging around this negatively charged core the way a comet swings around the sun.

But Rutherford hesitated to publish his conclusion. One reason was that it ran counter to the plum-pudding model of his mentor, J. J. Thomson, who was after all the world's leading expert in atomic physics. Rutherford then had a stroke of luck. One of J. J. Thomson's students, J. G. Crowther, published an experiment with beta particles that claimed to demonstrate that "the positive electricity within the atom . . . is distributed fairly uniformly throughout the atom."[14] This freed Rutherford from the Oedipal situation of having to attack his mentor directly; he was able to enter the fray by pouring forth scorn on Crowther and his conclusions while keeping warm relations with Thomson.

In an informal talk given in Manchester in March 1911, Rutherford referred to Crowther's results and conclusion—but then pointed out that Geiger and Marsden's discovery of large-angle scattering "cannot be explained" by the theory of multiple scattering. Instead, he said, "it seems certain that these large deviations of the alpha particle are produced by a single atomic encounter." This implied, in turn, an atom "which consists of a central electric charge concentrated at a point." Rutherford went on to bury Crowther's conclusion entirely by noting that his model could explain most of Crowther's experimental results as well.[15]

That May, Rutherford submitted to a scientific journal a "beautiful and famous paper," as Heilbron describes it. Its title is "The Scattering of α and β Particles by Matter and the Structure of the Atom."[16] After describing Geiger and Marsden's work, the theory of single and multiple scattering, and Crowther's experiment, Rutherford devoted a section to "General Considerations." In this formal presentation he wrote: "Considering the evidence as a whole, it seems simplest to suppose that the atom contains a central charge distributed through a very small volume." One of the seminal scientific papers of all time, it brought about, Rutherford's associate E. N. da C. Andrade said, "the greatest change in our ideas of matter since the time of Democritus . . . four hundred years before Christ." Atoms were supposed to be the basic building blocks of matter—the word "atom" is from the Greek for "uncuttable"—and here was a description of its inner parts and structure.

By providing a picture of atomic structure, Rutherford's model opened the door to solving many of the problems of atomic physics. Alpha particles, for instance, were indeed pieces of the nucleus that somehow had been ejected or chipped off—and were positively charged, like the rest of the nucleus, until they had slowed down enough to attract electrons, whereupon they became electrically neutral, like ordinary helium atoms.

Nevertheless, neither Rutherford nor anyone else at the time seems to have viewed this discovery as incredible or epoch-making.

Rutherford did not crow about the discovery in his correspondence, and he made only two brief references to this paper in a book he published almost two years later, *Radioactive Substances and Their Radiations*. The scientific world as a whole was equally quiet. There are virtually no references to Rutherford's paper in the leading scientific journals of the day, nor in the reports of key scientific conferences, nor in the lectures given by eminent scientists, including J. J. Thomson.

We in the twenty-first century, all too painfully aware of the subsequent and dramatic history of the nucleus, find this startling. But Rutherford's model was not yet connected with the massive amount of information that chemists and physicists knew about the atom. Indeed, his model, strictly speaking, could not have worked, since according to what was then known it was mechanically unstable. Only when Danish physicist Niels Bohr arrived in Manchester in 1912 and applied to Rutherford's model the idea of the quantum—that energy at the smallest levels does not come in any old amounts, but only in packets of certain specific sizes—did the model suddenly appear stable. Not only that, Bohr showed how the model, revised in the light of quantum theory, explained a lot more besides, such as the frequencies at which hydrogen atoms emit light. Still later, another Rutherford student, Harry Moseley, demonstrated that the Rutherford-Bohr atom accounted for the frequencies at which the innermost electrons of elements emit X rays. Only then would the nuclear atom become as obvious to others whose physical intuition was not as strong as Rutherford's.

Today, it is easy to describe Rutherford's experiment retrospectively, as he once did, as if its discovery were a "eureka" moment. Physics textbooks have likened the experiment to the way early customs inspectors would search for contraband in shipments of hay bales by shooting bullets through them; if the bullets ricocheted, the inspectors knew they had hit something inside the bales much denser than hay. But when Rutherford and his assistants embarked on this experiment, it was not clear that alpha particles were like bullets, nor was it clear what made them ricochet, or how. All these things emerged while the experiment was being born, not before. And only long after its conclusion was it obvious just how epochal a discovery Rutherford and his team had made.

ARTISTRY IN SCIENCE

I ONCE CONCOCTED A SCHEME to repeat the experiment by which Rutherford discovered the atomic nucleus. In practice, it looked simple enough: a source, a target, scintillation screens, tiny flashes that you counted in the dark. I had taken the trouble to collect pictures and diagrams of the experiment, accounts written by the various participants, and analyses by historians of science. I had even boned up on the math. I envisioned performing the experiment in front of students—making a video or documentary, perhaps. To assist me, I approached someone I knew had worked with Rutherford on alpha-scattering experiments, the physicist Samuel Devons at Barnard College. I went to his office to outline the scheme.

My suggestion made Devons literally roar with laughter—for a long time. After he finally quieted down, he explained to me that obtaining the permits to work with radioactive materials of the requisite strength was virtually impossible today. One could cheat, he said, and do what college laboratories sometimes do—use permissible weak sources with modern-day electronic equipment that one could leave alone for hours or days to collect the data. But that was evidently not what I had in mind. Then he said:

> The main problem, though, is that experiment is a craft,
> like making an old violin. A violin isn't a very complicated-

looking gadget. Suppose you went to a violin maker and said, "Could you kindly help me make a Stradivarius? I'm interested in violin making and I'd like to see how it was done." He'd smile at you just like I did. Because craft is a knowledge you have in your fingertips, little tricks you learn from doing things, and they don't work and you do them again. You have little setbacks, and you think, how can I overcome them? And then you find a way. Every time your experiment changes, you forget all the old techniques and have to learn new ones. And you have to know them, because when you're pushing your equipment to the limit, it's bloody easy to get spurious results. You're scratching at the ground all the time, and you don't know what you've missed. Every experimenter has made terrible errors at one time or another, and knows of instances where friends have fallen on their faces because they got spurious results and published too early. And yet you've got to push what you know to the limit. If you don't, someone else is going to do it first. And that's dreadful, being beaten. Everyone's got a closet full of discoveries they missed because they were too cautious or some other fellow was cleverer. There was a whole Austrian school working on the same things as Rutherford at about the same time, and nobody's heard of them today. Why not? Rutherford was just a little more daring and crafty.[1]

The kind of craft knowledge that Devons described is not found, of course, in physics alone. Albert E. Whitford, an influential mid-century American astronomer, remarked that, in his time, using a large telescope demanded "high artistry—doing it yourself. Real mastery of a beautiful and cantankerous instrument, a big telescope." And indeed, learning the intricacies of the machine was challenging work. "Observing at a telescope, even under the best of conditions, is tedious," Allan R. Sandage, an influential cosmologist who spent countless nights taking data with large telescopes, has remarked. "Under the worst, it can be cold and miserable." Yet the long, uncomfortable hours alone with the telescope under the night sky also fostered what science historian Patrick McCray called "an intimate bond between scientist and machine"[2]—the deep understanding necessary for the experimenter to know what the instrument is revealing and what it is not.

When such a bond exists, the result is performance that may be called artistic.[3] Performances can be classed in three groups: mechanical repetitions, standardized performances, and artistic performances. Mechanical repetition is exemplified by CDs or player pianos, which are encoded with signals that cause a device to recreate a piece of music. But the music, no matter how beautiful, is not a creation; it is only the echo of one. Standardized performances, by contrast, involve a minimum of artistry; actions that can be executed only by a few trained people become transformed into a practice that a much wider group of people without as much training can execute successfully, like the surgical technique for using lasers to restore the vision of the nearsighted. Once it was the province of costly specialists; now it's performed by chains of commercial medical clinics.

Artistic performance goes beyond the standardized program; it is action at the limit of the already controlled and understood; it is risk. As Rutherford's discovery of the atomic nucleus reveals, scientific objects have to be brought into focus from out of an often confusing background. The process can be likened to the experience of studying an optical illusion in which the outline of some object is hidden in a complex drawing. At first the object's features are mixed confusingly amid a tangle of lines and shapes, producing a vague tension and unease, until suddenly our sight is recast and we see the object—a rabbit, say—amid a thicket of leaves, sticks, and grass. Scientific objects are often recognized via an analogous process. In the laboratory, however, we are never sure that an object is really present to begin with. Furthermore, our instruments produce the drawing—both the object and the background in which we have to make it out. As a result, how we stage the experiment may interfere with our ability to recognize a new phenomenon, and we may have to alter the experiment before what we are seeking comes into view.

Rutherford's experiment illustrated not just artistry at work, but also how artistry becomes standardized and transformed into technique. Scientific phenomena can follow a trajectory from newly discovered effect (even a nuisance, in the case of Rutherford scattering) to laboratory technique and finally to technology. An effect is some characteristic, instructive, or useful consequence of a scientific phenomenon; alpha-particle scattering is one example. When an effect is sensitive to some sought-after parameters of a system— as Rutherford's alpha scattering was to charge and mass distribution—it can be turned into a technique because the effect can be used to alter, analyze, or measure those parameters. And it is always possible that the technique can mutate further into a technology— that is, become standardized enough to be performed by commercially available "black box" instrumentation whose principles do not have to be fully grasped by the user. Consider piezoelectricity, the phenomenon in which certain crystals, many naturally occurring, produce momentary jolts of tens of thousands of volts of electricity when squeezed in the right way. The phenomenon first put in a laboratory appearance around the turn of the century; it was made manifest by the artistic work of the Curie brothers, who produced it with complicated laboratory equipment (one of the brothers, Pierre, subsequently married Marie Curie, who went on to become the first woman to win a Nobel Prize). By World War II, piezoelectricity had been sufficiently standardized to use in the detonators of aerial bombs. Standardized even further, this once-exotic laboratory phenomenon is today a commonplace feature of the ignition systems of certain kinds of cigarette lighters.

Why, then, is the artistry of experiment so often overlooked? One reason is the attitude of scientists themselves, who often hold themselves and their colleagues to an exacting, unsentimental, and ultimately even unrealistic standard. Nobel laureate Leon Lederman, for instance, the former director of Fermilab, the national laboratory in Batavia, Illinois, had frequently chided himself over his "missed discoveries," and once wrote a paper about what he later called "the big ones that got away." Lederman counted among these the time when his team just missed spotting an important particle that six years later was simultaneously discovered by two other teams of researchers. "Our thinking," Lederman wrote, "[and] our grasp of the crucial elements of the physics, were fuzzy." But the work of Lederman's team was treated by colleagues as first-rate; indeed, both teams that finally discovered the particle, now called the "J/psi," used his earlier work as a guide. When I met Lederman, I asked him whether he really believed that he had lacked a sound grasp of physics in that experiment. "It wasn't sound enough," he replied. "But the experiment was, and still is, regarded by your colleagues as wonderful," I said. "Not wonderful enough," he replied. "If it had been a little more wonderful, we would have found the J/psi. I should have been smart enough to use fine-grained detectors." When I reminded him that he had used thick materials that precluded the use of that kind of detector, Lederman shook his head obstinately. "I should have been smart enough to take out the thick materials and put in thinner ones." "But," I protested, "that would have meant changing the entire scientific goal and physical structure of the experiment on highly speculative grounds." Lederman was unmoved. "If I had been smarter," he brooded, "I would have started that experiment over from scratch. But I wasn't. I was dumb."[4]

Why do Lederman and other scientists habitually adopt a self-deprecating attitude about their efforts, and refuse to acknowledge the artistry, and thus potential fallibility, in it? Their attitude, a convention that defines what it is to have "the right stuff" in science, attributes all failure to poor planning and judgment, and denies the inherent risks and uncertainty in experimental efforts. This attitude inspires them to greater effort in their risky and demanding work.

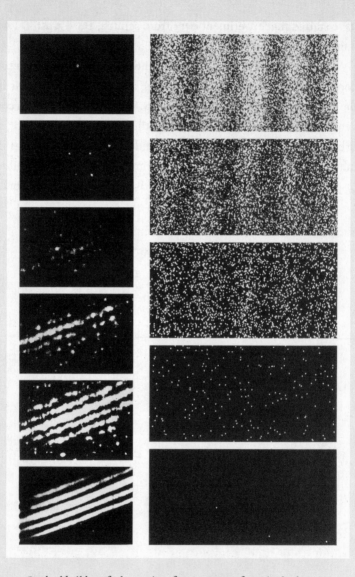

Gradual buildup of electron interference pattern from single electrons, from the Bologna group's 1974 experiment (left), and the Hitachi group's 1989 experiment (right). The Bologna group's normally vertical lines were rotated by a magnetic lens in the electron microscope.

10

THE ONLY MYSTERY:

The Quantum Interference of Single Electrons

—

*We choose to examine a phenomenon which is impossible,
absolutely* impossible, *to explain in any classical
way, and which has in it the heart of quantum mechanics. In reality, it contains the* only *mystery.*

—RICHARD FEYNMAN

"I SAW IT DURING AN OPTICS COURSE AT EDINBURGH University," wrote an astronomer in response to my *Physics World* poll, referring to the two-slit experiment with electrons. "The prof didn't tell us what was going to happen," she continued, "and the impact was tremendous. I cannot remember the experimental details any more—I just remember the distribution of points that I suddenly saw were arranged in an interference pattern. It is utterly arresting in the way a masterpiece of art or sculpture is arresting. Seeing the two-slit experiment done is like watching a total solar eclipse for the first time: a primitive thrill passes through you and the little hairs on your arms stand up. *Christ,* you think, *this particle-wave thing is really true,* and the foundations of your knowledge shift and sway."

In his *Lectures on Physics,* the late American physicist and Nobel laureate Richard Feynman remarked that "things on a very small scale behave like nothing that you have any direct experience about." Nonetheless, as Feynman well knew, it is all too easy for even the most knowledgeable physicist to ignore the complications of quantum mechanics and, despite their deep understanding of the subject, to imagine that electrons, protons, neutrons, and other particles "down there" are just like bodies "up here"—that is, solid, individual objects that take definite paths when they travel from point A to point B, and that if for some reason we lost track of them in between, they were still "there" in one place at one time. But we can set up experiments to show that this is not what happens in the quantum realm. This runs dead against the assumption, a firmly held part of science since Eratosthenes' experiment helped us picture the heavens, that we can somehow imagine or picture fundamental things.

The single most visible and dramatic demonstration that we cannot do so—that the activities of the quantum world cannot be pictured—is a version of the two-slit experiment performed by Thomas Young, but this time using not light but subatomic particles such as electrons. Because of the technical difficulty of preparing this experiment, and because it was developed in stages, it alone of the ten most beautiful experiments is not associated with a single name. It is simply referred to as the two-slit experiment, or quantum interference experiment, with single electrons. According to my survey, this was far and away the most often mentioned. My poll, to be sure, was unscientific. But have little doubt that the simplicity, undeniability, and shock value of the two-slit experiment would make it place high on *any* list of the most beautiful scientific experiments.

IN HIS LECTURE COURSE and in other books, Feynman elegantly described the weird nature of quantum behavior by com-

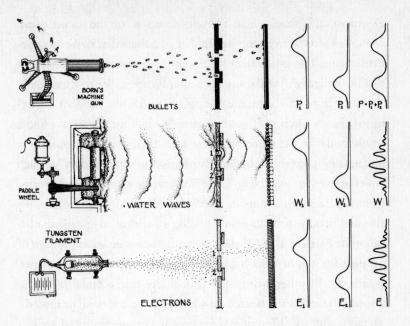

Figure 10.1. Three two-slit experiments, illustrating no interference with
"lumpy" objects (bullets), interference with continuous objects (water waves),
and interference with seemingly "lumpy" objects (electrons).

paring a trio of two-slit experiments—one using bullets (particles),
one using water (waves), and one using electrons (both and neither)—to progressively work out "by a mixture of analogy and
contrast" the similarities and differences in each case.[1]

First, Feynman said, imagine an experiment in which a machine
gun randomly sprays bullets over an armor-plated wall with two
small holes in it. Each hole has a shutter that can be used to close it
completely. Each hole is just big enough to let a bullet shoot through
it and strike the backstop. All but a few of the bullets hit the backstop in the same two places; a small number ricochet off the edges of
the holes and go off at an angle, so that we can never predict exactly
where any particular bullet will land. As part of this experiment,

Feynman said, imagine that a "bullet detector" on the backstop can be moved around to count the number of bullets that strike any particular spot. The experiment's purpose is to measure the probability that the bullets will strike any particular location. When we actually shoot the guns and begin measuring, we find first of all that the detector always catches, of course, complete bullets: We always find a whole bullet in the detector, never half a bullet or a fraction of a bullet. The pattern of bullet distribution is thus "lumpy"—each bullet arrives in "one bang," Feynman wrote—and each measurement is of a specific number of whole bullets. Most important, we also find that the probability of finding a bullet at any particular location when Hole 1 and Hole 2 are both open is equal to the sum of the probabilities of what happens when Hole 1 and Hole 2 are open separately. In other words, the probability that a bullet will pass through Hole 1 is unaffected by whether Hole 2 is open or closed.[2] To put it slightly differently, if you are at a firing range and are hitting the bull's-eye a certain percentage of the time, that percentage does not change even when someone sets up a second target nearby and begins to miss or hit your target. Feynman calls this a condition of "no interference."

Now, Feynman said, imagine a second experiment, this one involving a water tank and wave machine instead of a machine gun. This experiment, too, has a wall with two holes in it, an absorbing backstop or "beach" on the other side that does not reflect the waves striking it, and a moveable detector that measures the intensity of the wave motion (actually, it measures the height or amplitude of the wave and squares that number to yield the intensity). This is essentially Young's double-slit experiment applied to water waves.

The goal of the experiment is to measure the intensity of the wave motion when Hole 1 and Hole 2 are open separately and together. When we set the wave machine going, Feynman said, we spot several key differences from the previous experiment. First, waves can be any size—they are not lumps, like bullets—and their height can vary smoothly and continuously. Furthermore, the pattern of intensity variation when both holes are open is not the same as the sum of the patterns when each is open singly. The reason, as we know from Young's experiment, is that the waves from the two sources are in phase at some places and out of phase at others. Here we have a condition of "interference."

Finally, Feynman's third imagined experiment used an electron gun that shoots a beam of electrons against a wall with two holes in it. Again, on the other side of the wall are a backstop and an electron detector. We are here dealing with quantum behavior, Feynman said, and something very peculiar happens. As in the first experiment, we detect a "lumpy" distribution pattern, since the electrons seem to arrive at the detector singly and completely—either the detector gives off a "click," a machine-made event registering the arrival of an electron, or it does not. But as in the second experiment, the pattern of electron distribution when both holes are open is not the same as the sum of the patterns when each is open singly. The result is a classic interference pattern. Astoundingly, the electrons acted like waves while passing through the slits, but like particles when activating the detectors.

You might imagine that, since many electrons pass through the two slits at once, the interference pattern arises somehow because of the fact that many of them collide with one another. But a variation of the experiment, involving one electron at a time, shows that this is not so. Here we reach the "only mystery."

Let us then turn down the electron gun so that it spits out only one electron at a time, at a slow enough rate that there is never more than one electron at a time passing through the slit. Now it is impossible for collisions between electrons to occur. When we switch on the electron gun, the electrons slowly accumulate on the other side. At first, as the electrons collect at the detector, they seem to show up randomly. But as the data grows, we are startled to see a pattern forming—in fact, an interference pattern! Apparently, each electron moves through both slits at once, like a wave, but hits the detector at a single point, like a particle. Each electron interferes only with itself. Can this be true? It is—and this is the "only mystery," Feynman said. "I am avoiding nothing; I am baring nature in her most elegant and difficult form."

Because single electrons are difficult to produce and observe reliably in isolation in such a gun, physicists thought for a long time that it would be impossible actually to perform this experiment. Still, they were utterly confident about what would happen if it could be performed, because they had seen so much other evidence of the wavelike nature of electrons. As Feynman told his students:

> We should say right away that you should not try to set up this experiment. . . . This experiment has never been done in just this way. The trouble is that the apparatus would have to be made on an impossibly small scale to show the effects we are interested in. We are doing a "thought experiment," which we have chosen because it is easy to think about. We know the results that *would* be obtained because there *are* many experiments that have been done, in which the scale and the proportions have been chosen to show the effects we shall describe.

When Feynman said this, in the early 1960s, he was apparently unaware that technology was advancing to the point where experimenters could set up a real quantum two-slit experiment. In fact, it had already been done, in 1961, by a German graduate student named Claus Jönsson.

Born in Germany in 1930, Jönsson was young enough not to be conscripted during World War II. When the Allies drove the German army back through Hamburg, Jönssen and a group of science-minded high school classmates picked through equipment abandoned by the German troops. They stripped a German jeep of its battery and other electrical parts, and conducted electroplating experiments. Their fun came to an end only when, having no access to recharging equipment, they ran down the battery.

After the war, Jönsson studied at the University of Tübingen under Gottfried Möllenstedt, a pioneer in electron microscopy, who worked at the Physical Institute at the university.[3] Möllenstedt was the co-inventor (with Heinrich Düker) of the electron biprism, which is essentially a Fresnel biprism for electrons (Figure 10.2). As I described in Chapter Six, Young's two-slit device and Fresnel's biprism were two different but conceptually similar methods of splitting a light beam into two sets of waves that interfere with each other. Young's method divided light from a single source into emanations from two slits separated by a small distance; Fresnel divided light from a single source by having it pass through two sides of a triangular prism simultaneously. Möllenstedt's electron biprism effectively divided a beam of electrons into two components by placing an extremely thin wire across it at a right angle. The wire had to be so extraordinarily fine that at first Möllenstedt coated the silk strands from a spiderweb with gold (he kept a collection of spiders around the laboratory for this purpose). Later, he found a better, cheaper way to manufacture ultrathin wire by using quartz fibers stretched out in a gas flame and then gold-plated. When the biprism fiber was charged positively, it effectively split the beam into two components slightly tilted toward each other, allowing them to interfere.

In the summer of 1955, Möllenstedt and Düker called together the Institute's co-workers, including Jönsson, to show off the first interference patterns produced with the biprism. Soon after, Jönsson conceived the idea of replacing the biprism with a small double slit, in explicit parallel to Young's experiment, to try to create two-

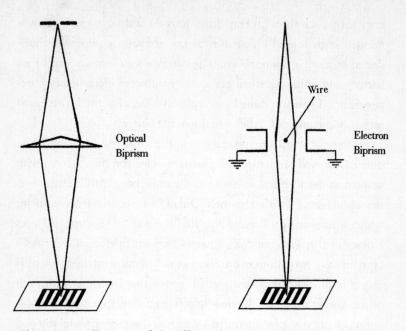

Figure 10.2. *Illustration of the difference between an optical Fresnel lens (left) and a Möllenstedt-Düker electron biprism (right).*

slit interference fringe patterns with electrons. The obstacles were formidable. He would have to cut extremely tiny slits into a special metal foil. While optical slits can be mounted on some transparent material, such as a glass slide, this would be impossible for the electron experiment, because every such material would scatter the electrons. The foil would therefore have to be mechanically stable, or strong enough to stand on its own without recoiling when struck by electrons. Here Jönsson encountered the experimenter's trade-off, since cutting slits in a substrate (supporting) material thick enough to absorb electrons tends to leave uneven edges, while making the material thinner makes the slits more precise but harms the ability of the material to support itself without wobbling, which would affect the behavior of the electrons passing through the slits.

The slits, meanwhile, would have to be much tinier than Young's, for the electron beam was only about ten-millionths of a meter (10 micrometers) wide. These slits would have to be clean, because the electrons would bounce off any irregularity and spread out randomly, destroying what is known as the "coherence" of the electrons. Here Jönsson's experiences with the German jeep battery came in handy, since they had left him with a deep appreciation of the importance of the cleanliness of the substrate. But several senior scientists strongly doubted whether Jönsson could succeed and insisted that he give up the idea. But with the encouragement of Möllenstedt—who informed him that "*'Es geht nicht' gibt es nicht für einen Experimentalphysiker*"["'It won't work' is not in the vocabulary of an experimental physicist"]—Jönsson went ahead anyway.

After finishing the first part of his doctoral examinations in 1956, Jönsson began to explore methods for cutting slits in foil of sufficient thinness, and by the following year he had found a way.[4] In spring 1957, Jönsson passed the final, theoretical part of his examinations, and approached Möllenstedt to talk about a dissertation topic. Möllenstedt had originally wanted Jönsson to work on biprism interference, but he agreed to let Jönsson change the subject. The first part of the project involved building a machine that could create slits less than eight hundred billionths of a meter (800 nanometers) in size—a device so far in advance of its time that Jönsson became one of the pioneers in what is now called "nanotechnology." The second part required him to develop a special film that would work even with the low intensity of the electrons.[5] A constant problem was to eliminate mechanical and magnetical disturbances that would distort the interference pattern. Jönsson got the first photograph of an interference fringe pattern in 1959 (Figure 10.3), and received his doctorate for the work in 1961.

Everyone versed in quantum mechanics knew that Jönsson's experiment broke no new theoretical ground, and nobody was sur-

Figure 10.3.
Electron
interference
pattern from
Jönsson's
experiment.

prised by the result. Still, he derived considerable satisfaction from bringing into reality what he later called "an old thought experiment of quantum mechanics which before had seemed to be impossible, and an experiment of high pedagogical and philosophical importance." And when his paper was translated and published in English, in the *American Journal of Physics*, a journal for physics educators, the editors went out of their way to praise Jönsson's experiment. While not on the cutting edge of theoretical physics, they wrote in their lead editorial, it was still a "great experiment" and "technical *tour de force*" that provided "the conceptual simplicity of a real, pedagogically-clean, fundamental experiment whose description and study can now enrich and simplify the learning of quantum physics." In so doing, it helped supply "the juices of experimental reality that . . . transform a formal discipline into a living profession."

At that time, it was still impossible to carry out the experiment with single electrons, but within about a decade this, too, changed. This final version of the two-slit experiment likewise originated in interesting circumstances. In 1970, Pier Giorgio Merli and Giulio Pozzi, two young researchers at the Electron Microscopy Laboratory at the University of Bologna, in Italy, attended an international workshop on electron microscopy in Erice, Sicily. Merli and Pozzi were especially impressed by a talk about new image intensifiers (essentially, light amplifiers) sensitive enough to detect single electrons, and upon their return were anxious to begin research projects using them. Their laboratory had been promised funding by the chief national agency that financed scientific projects, the Consiglio Nazionale delle Ricerche (CNR), but the funds had become stalled in the slow-moving government bureaucracy. The next year, 1971, the laboratory administration sent Pozzi and a senior researcher, Gian Franco Missiroli, to CNR's headquarters in Rome to try to discover what the holdup was.

On the train ride down, the two tried to take their minds off the stress of the looming visit—the sort of bureaucratic confrontation that nearly every scientist loathes and feels ill prepared for—by talking about physics. Pozzi mentioned to Missiroli his interest in working with an electron biprism, and the two began talking about possible projects they could work on together. It was the start of a fruitful thirty-year collaboration. Missiroli was not only an inventive researcher, but was also keenly interested in turning his discoveries into simple, easily teachable lessons for students, which he would write up and publish. The two began collaborating on experiments at the end of 1971.[6]

By that time, Merli had left the laboratory to take a position as a researcher at the newly established Laboratory of Chemistry and Technology for Materials and Electronic Devices (LAMEL), but still was able to collaborate with Pozzi, Missiroli, and other researchers of the Electron Microscopy Laboratory. The three men built a biprism and mounted it in a Siemens electron microscope. When Merli discovered that an image intensifier able to detect single electrons had been installed in Milan, the three began planning an electron-interference experiment in which they would send one electron at a time through an electron biprism. The three moved to Milan to shoot the images, attaching the image intensifier to their electron microscope, and succeeded immediately in detecting an interference pattern.

They wrote up the experiment and published it, as Jönsson had, in the *American Journal of Physics*, hoping that, they wrote, "electron interference experiments will become more familiar to students."[7] But they grew even more ambitious, and—with the encouragement and support of two other LAMEL scientists—decided to try to put together a short film about their experiment for distribution to local schools and libraries. However, this proved harder than it seemed and cost much more than expected, and the three wound up spending most of their time on the text. Experimenters, not theorists, they found that they had to work very carefully to express things with precision.

The result was ingenious. As had Feynman and many others, they, too, used a three-step analogy to explain the experiment, starting with the interference of water waves (first in nature, then in a ripple tank), then moving to the interference of light using a Fresnel bisprism, and finally describing their electron biprism. The three acted in the movie, which Merli edited. He also cleverly selected the background music, using Vivaldi flute music to accompany the explanation for the classical sections (the interference of water and light) and contemporary atonal music to accompany the quantum segments. The film culminated in showing the quantum interference pattern being slowly built up out of the accumulation of single electrons. The effect was magnificent, and the film (viewable on the Web) won an award at the International Festival on Scientific Cinematography in Brussels in 1976.[8] "Even today, every time I see the movie, I find it awesome," Pozzi told me, a feeling shared by the others.

In 1989, Akira Tonomura, senior chief research scientist at the Advanced Research Laboratory at Hitachi Limited Japan, and a group of co-workers carried out an electron microscope experiment using a still more sophisticated and efficient electron-detection system. They, too, published their work in the *American Journal of Physics*.[9] And they also put together a movie that shows the buildup of an interference pattern from the gradual accumulation of single electrons in real time. Tonomura showed this movie at a talk at the Royal Institution, which is also available on the Web.[10] At one point during his talk, he sped up the video to show the interference pattern materialize—hauntingly—out of individual, apparently random specks, the way a galaxy might form before your eyes at dusk out of tiny stars, a pattern that is undeniable and that hints at the existence of deeper universal structures. As this was happening, Tonomura said:

> We have no choice but to accept a very strange conclusion: that electrons are detected one by one as particles, but [that] the whole ensemble manifests wave properties to form an interference pattern. Quantum mechanics tells us that we have to give up the [conventional] reality of the particle picture of electrons except at the instant that we detect them.

In more recent years, quantum interference has been demonstrated with particles other than electrons, including atoms and even molecules.

The two-slit experiment applied to electrons possesses the three key aspects of beautiful experiments. It is fundamental, exhibiting as it does the strange and counterintuitive behavior of matter at the smallest level. An electron leaves a source, then shows up at a detector some distance away. Between production and observation, where was it? The quantum interference experiment—whether using two slits or a biprism—shows the impossibility of conceiving a quantum object as having the same kind of presence in space and time that objects in our macro world do. "Where was it?" is a question we cannot ask; it was everywhere and nowhere. If Young's two-slit light experiment was a dramatic illustration of the need to make a paradigm shift from light as a particle to light as a wave, the two-slit experiment with single electrons is a dramatic illustration of another paradigm shift, from classical to quantum physics.

It is economical, because despite its revolutionary implications the equipment is now within our technological reach and the basic concepts are readily understandable. Furthermore, this experiment displays in a concise way what is so mysterious about quantum mechanics. All the other mysteries of quantum mechanics—such as those illustrated by Schrödinger's famous cat, Bell's inequalities, and experiments involving nonlocality—stem from the mystery of quantum interference.

And it is convincing and deeply satisfying, able to convince the most die-hard skeptic of the truth of quantum mechanics. Even for someone well versed in quantum mechanics, the theory can be abstract, and its implications remote from our perception. But the two-slit experiment turns theory into an immediately graspable, sensuous image. "Before seeing it [in college], I didn't believe a single word of 'modern' [twentieth-century] physics," one scientist wrote me in my poll.

This experiment has something of the lucid beauty of Young's experiment, thanks to the immediacy of the evidence of the interference pattern. It has something of the beauty of expected surprise of the Leaning Tower experiment, which delights us by perceptual displays of the expectations of our everyday framework being violated; what this experiment shows, of course, is not mysterious if you are not accustomed to the idea that matter comes in discrete particles.

Finally, this experiment is beautiful—to me, anyway—for the way it effectively acts as a bookend to the astonishing feat of Eratosthenes. Eratosthenes' experiment validated the Greek intuition that the heavens possess an ultimate and picturable cosmic architecture; that on the biggest scale the universe consists of bodies moving about one another in three-dimensional space. The quantum interference experiment demonstrates that, on the smallest scale, things are interconnected in a way that cannot be conventionally intuited or pictured. Using equipment we have built with our own hands, we can see convincing evidence of an entirely different kind of world.

The quantum mechanical world is likely to remain counterintuitive to human beings, no matter how convinced we are of the theory. The double-slit electron interference experiment brings its reality before our eyes in a dramatic, economical, and material way. The experience of seeing the clicks of a detector tracking single electrons through a biprism or pair of slits and producing an interference pattern is one of the most awesome and arresting human experiences. The quantum interference experiment with single electrons is therefore likely to remain in the pantheon of beautiful experiments for a long time to come.

RUNNERS-UP

THE LIST OF RUNNERS-UP to my poll of the most beautiful science experiments consists of dozens of experiments in many fields. A few in particular are worth mentioning because of their circumstances, the unusual ways they manifest their beauty, or the fact that they are personal favorites.

The earliest of these runners-up was a hydrostatics experiment carried out—inadvertently—by Archimedes of Syracuse, the best-known mathematician and inventor of ancient Greece and, as it happens, Eratosthenes' contemporary. Today's science historians think it indeed plausible that, in the third century B.C., King Hieron of Syracuse asked Archimedes the proportion of gold and silver in a gift he had received. According to our ancient source for the episode, Archimedes was pondering the problem while sitting down in a bathtub, and noticed that "the amount of water which flowed over by the tub was equal to the amount by which his body was immersed [and this] indicated to him a method of solving the problem."[1] More likely—for measuring volume precisely through water displacement would be extremely difficult—what Archimedes realized was that he was being buoyed up (as would be the king's gift), and that if he could weigh it in air and water and compare the two, he could find the specific density of the crown accurately enough to compare it with gold. Did Archimedes then run naked through the town, yelling with joy? Perhaps not, though the story nicely captures the celebratory spirit that accompanies discovery. Still, the legend illustrates exactly how an inadvertent discovery can transform a routine event into a beautiful experiment.

Strong contenders in the life sciences included the so-called Meselson-Stahl experiment, the subject of a book by the late science historian Frederic Holmes entitled *Meselson, Stahl, and the Replication of DNA: A History of "The Most Beautiful Experiment in Biology."*[2] That experiment, carried out in 1957, confirmed that DNA replicates in the way predicted by the then recently discovered double-helix structure. Holmes took his subtitle from one researcher's description, but he noted that in fact most biologists familiar with this experiment felt the same way about it. When he asked the scientists why, their answers included simplicity, precision, cleanness, and strategic importance.

Contenders from psychology included two that simply but convincingly overthrew well-established dogmas of animal behavior. One, by American psychologist Harry Harlow, challenged the idea that the need for food was the most powerful factor in the bond between an infant primate and its mother. Harlow created a set of surrogate "mother monkeys," some made out of wire with no soft surfaces, others out of soft cloth. In one series of experiments, Harlow found that infant monkeys strongly preferred the cloth surrogate mother—even when the wire surrogate was the one whose nipple produced milk.[3] Evidently, the need for interpersonal bonding—for love and affection, as represented by softness—was more powerful than the need for food.

Another beautiful animal psychology experiment, by John Garcia and Robert Koelling in 1966, challenged B. F. Skinner's so-called equipotentiality laws of learning behavior, according to which an animal learns by stimulus and response, with conditioning able to connect any stimulus equally well with any response. Rats, for instance, could be taught to avoid a certain kind of flavored water by giving them electric shocks after they drank it. Garcia and Koelling repeated this lesson with one group of rats—but with another group of rats changed the stimulus and made them nauseated by the water instead. That worked faster and far more effectively than electric shocks. The experiment demonstrated convincingly that being sick and being frightened have very different effects on learning—and on the way animals interpret their environments. Yet it ran so counter to the then strongly held behaviorist doctrine of equipotentiality that Garcia's papers were rejected by American Psychological Association journals for more than a dozen years afterward.[4]

One engineering demonstration, beautiful for its importance, economy, and decisiveness, was Richard Feynman's famous act of dipping an O-ring in a glass of ice water during the investigation of the *Challenger* space shuttle disaster. He thus vividly demonstrated how its loss of resilience was the cause of the tragedy.[5]

Another intriguingly beautiful runner-up was the 1919 British expeditions demonstrating the gravitational bending of starlight, an epoch-making experiment that confirmed Einstein's 1915 prediction of this in his theory of general relativity and made Einstein a household name. But neither the eclipse (a familiar natural event) that made it possible nor the determination of stellar positions (a familiar astronomical technique) were extraordinary. Can beauty lie solely in an experiment's dramatic consequences?

Some theoretical arguments are so succinct that respondents referred to them as "beautiful." One is Stephen Hawking's proof, paraphrasable in fourteen words, that the universe has not existed forever ("It's true because if it were not, all things would be the same temperature"), and Olbers's paradox ("Look at the sky. It is not uniformly bright. Thus the visible universe is not infinite."). Some respondents cited certain experiments that used little beyond ingenuity to open up vast new domains for exploration. These included the Wilson cloud chamber (mentioned in Chapter Nine), which makes the tracks of charged particles visible—and that Ernest Rutherford once described as "the most wonderful experiment in the world." Other instruments that respondents cited included the X-ray interferometer, the scanning tunneling microscope, and the Cosmotron, a particle accelerator at Brookhaven National Laboratory.

An experiment whose beauty derives from the dedication of its creators was conducted by the Italian scientists Marcello Conversi and Oreste Piccioni during the Allied bombardment of Rome in World War II. Many Italian physicists had dispersed or even fled the country, but Conversi had avoided being drafted because of the poor eyesight in his left eye, while Piccioni was drafted but stationed in Rome. Prior to the Allied invasion of Sicily in July 1943, the two worked nights at the university, assembling stolen wire and radio equipment bartered for on the black market to create state-of-the-art electronic circuits with which they hoped to measure the lifetime of a puzzling particle, the mesotron, found in cosmic rays, the particles from space that continually bombard the earth's surface. After the invasion, American bombers began pounding the San Lorenzo freight station, located next to the university, and bombs occasionally struck the campus. Terrified, Conversi and Piccioni carried their equipment to a deserted high school close to the Vatican, which was spared bombing, though they had to share the basement with antifascist Resistance members who were storing weapons there. Conditions grew still worse after the Italian government signed an armistice with the Allies in September and the Nazis occupied Rome. Piccioni was once captured by German soldiers, but ransomed himself for a pile of silk stockings. The two continued to work feverishly—"Our work was the only pleasure we had," Piccioni once explained. Just before the Allies liberated Rome in June 1944, in an ingenious, elegant, and utterly convincing experiment, Conversi and Piccioni were able to show that mesotrons lived slightly more than 2.2 microseconds, a short lifetime but one that was still many times longer than predicted. In their basement in the ruined city, Conversi and Piccioni were the first to realize that mesotrons—now known as "muons"—had quite different properties from what the prevailing theory said, a key step in the emerging field of elementary particle physics.[6]

My own candidates for beautiful experiments would include the 1956–57 parity-violation experiment whose leaders included Chien-Shiung Wu. This experiment showed that under certain conditions particles and nuclei decay by emitting electrons in certain preferred directions with respect to their axis of spin. This experiment overturned at one convincing stroke one of the most fundamental and firmly held assumptions in physics.[7] I would also include Maurice Goldhaber's 1957 experiment establishing neutrino helicity, meaning the way that neutrinos spin with respect to their direction of motion. Goldhaber's experiment was so fiendishly ingenious—it involved coming up with a complex nuclear reaction in which the properties of all the particles and nuclear states involved were known *except* for the neutrino helicity, which was only possible in a single one of the three thousand or so known reactions—that most physicists at the time did not even think it was possible in principle.[8] While in most scientific discoveries one feels that, had the actual discoverers missed the boat, the discoveries still would have been made eventually, this one is different. One physicist later wrote that, had Maurice Goldhaber not existed, "I am not sure that the helicity of the neutrino would ever have been measured."[9]

But for my personal favorite candidate for most beautiful science experiment, read on.

Conclusion

Can Science Still Be Beautiful?

—

ALMOST ALL THE EXPERIMENTS IN THE TOP TEN WERE done solo, or with the aid of a few collaborators, in a relatively short period of time. But the past half-century has wrought tremendous changes in the size and scale of science experiments. Today, physics experiments are routinely interdisciplinary and multinational, and often involve dozens of institutions and hundreds of collaborators; they may take years or even decades to complete. Can an experiment in the age of Big Science still be beautiful?

Yes.

My personal candidate for the most beautiful science experiment, the muon g-2 experiment, has been done four times over the past half-century by ever-bigger collaborations, the first three times at the CERN international laboratory in Geneva and most recently at Brookhaven National Laboratory. In its latest incarnation, the collaboration swept up over a hundred scientists from several countries, who built a piece of equipment that included the largest superconducting coil in the world, inside a room the size of a small aircraft hangar. I should immediately confess that part of my affection for this experiment is purely personal. It is taking place in a building near me, and I've been watching it come together for years. But my familiarity with the experiment—just as with a particular complex novel or piece of music—merely deepens my appreciation of its beauty.

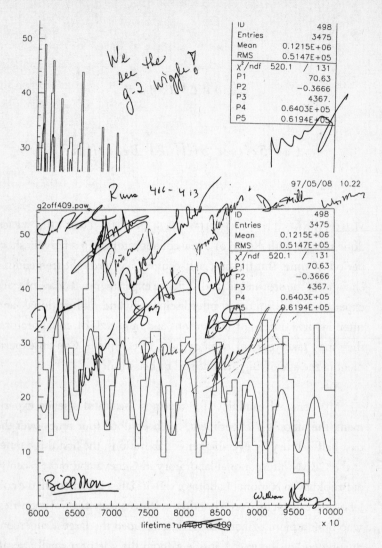

This plot of high-energy positrons as a function of time was the first evidence of stored muons precessing in the Brookhaven g-2 storage ring. Those present were so excited that they all signed it.

The experiment measures what is known as the "anomalous magnetic moment of the muon." It measures the way in which that particle—whose lifetime was first measured by Conversi and Piccioni—"wobbles" in a magnetic field.[1] Measuring that wobble required extraordinary precision, which in turn required devising an ingenious experiment.[2] To measure how muons wobble, the scientists study the electrons and positrons produced when they decay, making use of the phenomenon of parity violation discovered by Wu and her colleagues, which reveals the direction of the muons' spin.[3] When the data from the decay of billions of muons is plotted, it yields a dramatic-looking wiggle—a series of peaks, gradually declining in height, which reflect the frequency with which the muons are wobbling inside the chamber.

The first data of the most recent Brookhaven experiment came through in May 1997. The first scientist to add a few days' worth of data together and plot them on a graph paper, a University of Minnesota physicist named Priscilla Cushman, noticed the telltale wiggle immediately. When Gerry Bunce, another member of the g-2 team, entered the room, Cushman recalled,

> I thrust it under his nose and said, "Look! The g-2 wiggle!" He said, "We're having a party tonight!" I said, "But we have so much work to do!" But Gerry was right. We had been waiting so long, gone through such hard times with funding agencies and other critics who said we would never succeed—and then there were two weeks when we first turned on the machine when we saw nothing—and suddenly this *beautiful thing* just popped out and we saw g-2 was there!

Several years of data collection later, the experimenters had taken one of the most precise measurements of a property in particle physics ever performed, and were able to compare it to the theoretical value, which is one of the most precisely calculated—and measured—numbers in science.[4] The results indicated a discrepancy with the number predicated by theory, suggesting that new physics might lie on the horizon and creating great excitement among physicists.

The g-2 experiment exhibits the three elements of beauty we have encountered in the other experiments discussed in this book: depth, or how fundamental the result is; efficiency, or the economy embodied in the parts; and definitiveness—that if questions are raised, they are more about the world (or the theory) than about the experiment itself. And despite its scale, the g-2 experiment has the breadth of Eratosthenes' experiment, linking different scales of the universe (phenomena of vastly different energies) in one tiny measurement—that of the muon's wobble. It has the austere beauty of Cavendish's weighing-of-the-world experiment, where precision had to be fanatically pursued through myriad interconnected pieces. It has the synoptic quality of Millikan's experiment, for it brings together many different laws of the universe to achieve its result, from electromagnetism and quantum mechanics to relativity.[5] And it has something of the sublime beauty of Foucault's pendulum, hinting at yet-to-be-conceived dimensions of the universe.

IN THE INTRODUCTION I posed two questions about the idea of beautiful experiments. First, what does it mean for experiments, if they can be beautiful? Second, what does it mean for beauty, if experiments can possess it?

To answer the first question: Understanding how experiments are beautiful helps us to appreciate their affective power. Many respondents to my poll mentioned experiments and demonstrations they had seen as schoolchildren—indeed, often these were the *only* things that they recalled from their early science education. Looking at the moon through a telescope for the first time, peering through a microscope at the pulsing veins in a goldfish's fins, holding a spinning bicycle wheel by the axle and feeling the resistance when trying to turn it over, seeing a beach ball float in a strong vertical air current, watching a steel can crumble when the air inside is evacuated—events like these have an uncanny ability to hold our imaginations.

Experiments enthrall not just students but also seasoned scientists. The thrill of discovery is like no other—which is why the Scottish engineer John Scott Russell reacted to his sighting of a soliton wave (an isolated wave that does not disperse, as do ordinary waves) in Edinburgh's Union Canal in 1834 by calling it "the happiest day of my life." Similar experiences abound in the history of science.

Historians and philosophers all too often ignore the passions clearly evident in such stories. Some scholars do so to emphasize the rationality of science; its logic or justification. But the picture that emerges suggests that science is a robotic process of hypothesis formation, testing, and hypothesis reformulation—a vast intellectual game. On another front, some historians and philosophers explore the social dimensions of science, its social context as revealed by its politics, funding, or benefits.[6] These are indeed interesting subjects, but these accounts tend to suggest that science is merely a vast power struggle conducted by a special interest group determined to advance its own cause.[7] If we know science only by its logic or justifications on the one hand, or by its politics, interests, funding, or material achievements on the other, then we fail to understand it. If we take the time to examine the beauty of science experiments, we can bring that affective dimension back into focus.

The answer to the second question raised above is that recognizing the beauty of experiments can help revitalize a more traditional sense of beauty. The term is applied most commonly nowadays to artworks and natural phenomena, but this was not always the case—and if we know beauty only by sunsets or the contents of art museums, we fail to understand its full role in human life and culture. The ancient Greeks saw no special connection between the beautiful—what they called *to kalon*—and artworks, viewing the beautiful as anything valuable or worth seeing in and of itself. They associated beauty not with decorations or flourishes, but with exemplary things, including laws, institutions, souls, and actions. As a result, they perceived an intimate connection between the true, the beautiful, and the good, seeing these as "entangled," so to speak, inseparably bound up together in a common and deeply lying origin.

Plato called beauty the shining of the ideal in the realm of the visible. Beauty is the glow that true and good things have—simultaneously enlightening, compelling, and satisfying—when they appear in the world inhabited and perceived by finite humans. The higher orders of nature announce themselves to lovers of wisdom by their beauty. For this reason, Plato held, the lover of knowledge does not neglect, but carefully cultivates, a sense of the beautiful—for to do so is at the same time to cultivate a sense of truth. The world is never fully transparent to us, and we meet it with historically and culturally transmitted assumptions that both reveal and conceal much. But we also encounter things, which we call beautiful, that draw us out of our confusion and ignorance. Beautiful things, Plato writes in the *Symposium*, call us more deeply into the world; they are like "rising stairs" that take us "always upwards."[8] Stairs and transitions always take us from one location to another; the human place in the world is not fixed but mobile. And when we allow ourselves to be led upward, we achieve a more intimate connection with ourselves and the world, and thereby become more human. Thus the ability to recognize the beauty of experiments can help reawaken us to a more original and fundamental sense of beauty itself.

> *The scientist does not study nature because it is useful; he studies it because he delights in it, and he delights in it because it is beautiful. If nature were not beautiful, it would not be worth knowing, and if nature were not worth knowing, life would not be worth living.*
>
> —HENRI POINCARÉ

Acknowledgments

THIS BOOK GREW OUT OF AN ARTICLE I DID FOR *Physics World*, and I am greatly indebted to its editors, especially Matin Durrani and Peter Rodgers, for allowing me the opportunity to write a column for that magazine, as well as to the hundreds of people who responded to my original survey. I wrote this book (while pursuing several other projects) during part of my sabbatical from Stony Brook University at the Dibner Institute for the History of Science and Technology, at MIT, and am indebted to its director, George E. Smith, and to other members of its staff—Carla Chrisfield, Rita Dempsey, Bonnie Edwards, and Trudy Kontoff—as well as to staff members of the Burndy Library: Anne Battis, Howard Kennett, David McGee, Judith Nelson, and Ben Weiss. I am indebted to my literary agent, John Michel, who steered me imaginatively in the right direction, and to my editor, William Murphy. Like all columnists, I rely heavily on others for inspiration, ideas, and information, and those who provided helpful suggestions, comments, and other kind of assistance include: Philip Bradfield, Edward Casey, Elizabeth Cavicchi, Stephanie Crease, Robert DiSalle, Patrick Heelan, Jeff Horn, Thomas Humphrey, Don Ihde, Claus Jönsson, Kate from South Country, Jean-Marc Lévy Leblond, Gerald Lucas, Peter Manchester, Alberto Martinez, Pier Giorgio Merli, Lee Miller, Arthur Molella, Giulio Pozzi, Patri Pugliese, Evan Selinger, Thomas Settle, Steve Snyder, Bob Street, Clifford Swartz, Akira Tonomura, Jeb Weisman, Evan Welsh, Donn Welton, and many others. As always, I have been energized by the sounds of surprise. Finally, I want to thank Jack Train, Jr., whose innovative science writing, brilliant editing, and generosity have been an inspiration for decades.

Notes

INTRODUCTION:
THE MOMENT OF TRANSITION

1. For a nontechnical account of this experiment, see Robert P. Crease and Charles C. Mann, *The Second Creation: Makers of the Revolution in Twentieth-Century Physics*. New Brunswick, N.J.: Rutgers University Press, 1996, pp. 386–390.

2. Watson's remarks: Victor McElheny, *Watson and DNA: Making a Scientific Revolution*. Cambridge, Mass.: Perseus, 2003, p. 52. For Millikan's remarks, see Chapter Eight.

3. The first Weisskopf quote is taken from K. C. Cole, *The Universe and the Teacup: The Mathematics of Truth and Beauty*. New York: Harcourt Brace, 1998, p. 184 (the date is from a personal communication). The second Weisskopf quotation is from V. Stefan, ed.: *Physics and Society: Essays in Honor of Victor Frederick Weisskopf by the International Community of Physicists*. New York: Springer, 1998, p. 41.

4. G. H. Hardy, *A Mathematician's Apology*. Cambridge, Mass.: Cambridge University Press, 1992, sections 10–18. On beautiful equations, see G. Farmelo, ed., *It Must Be Beautiful: Great Equations of Modern Science*. London: Granta Books, 2003.

5. Michael Faraday, *The Chemical History of the Candle*. New York: Viking, 1963, Lecture 1.

6. Whether we portray this process as the world speaking to us, as in the classical view, or our own projected words returning to us, as the social constructivists would have it, is entirely irrelevant. The essential point is that experimentation is a meaning-producing event that is more complex than is able to be portrayed by these two accounts. See Robert P. Crease, "Hermeneutics and the Natural Sciences: Introduction," in Robert P. Crease, ed., *Hermeneutics and the Natural Sciences*, Dordrecht: Kluwer, 1997, pp. 259–270.

7. Mark Twain, *The Innocents Abroad*. New York: Literary Classics of the United States, 1984, pp. 196–197.

8. Schiller's notion of beauty is discussed at length throughout his book *On the Aesthetic Education of Man*. Ralph Waldo Emerson, *Essays & Poems*. New York: Literary Classics of the United States, 1996, p. 931.

9. Robert P. Crease, "The Most Beautiful Experiment," *Physics World*, May 2002, p. 17; Robert P. Crease, "The Most Beautiful Experiment," *Physics World*, Sept. 2002, pp. 19–20.

10. Eratosthenes' experiment is part of the curriculum of numerous junior high school science classes, was discussed in Carl Sagan's popular *Cosmos* series, and is the subject of a children's book. Galileo's Leaning Tower experiment is practically as legendary as the tale of George Washington chopping down the cherry tree, and was reenacted on the moon by the crew of Apollo 15. Galileo's inclined-plane experiment is taught in many science classes and appeared in a scene of a Philip Glass opera, *Galileo Galilei*. The meaning of Isaac Newton's experiments with prisms was hotly debated by poets and writers throughout the eighteenth and nineteenth centuries. Foucault's pendulum has a stamp of cultural legitimacy because it appears in many public institutions—including the United Nations building in New York City—and was featured in at least two novels, including a best-seller by Umberto Eco entitled *Foucault's Pendulum*. Two of these experiments—Millikan's oil-drop experiment and Rutherford's experiment discovering the atomic nucleus—have been the subject of influential and controversial papers by historians of science. Impressive and widely circulated videos have been made by two independent teams of scientists who have conducted the two-slit experiment illustrating quantum interference with single electrons. Tom Stoppard's play *Hapgood* includes a discussion of this experiment, as well as Young's two-slit experiment with light. And so forth.

CHAPTER 1: MEASURING THE WORLD:
ERATOSTHENES' MEASUREMENT
OF THE EARTH'S CIRCUMFERENCE

1. Aristotle, *On the Heavens*, tr. J. L. Stocks, 298a. In *The Works of Aristotle*, Vol. 1, Chicago: Encyclopaedia Britannica, Inc., 1952.

2. *Ibid*.

3. These sources include: Cleomedes, Capella, Strabo, Pliny, Aelius Aristides, Heliodorus, Servius, and Macrobius. See the extracts in: A. S. Gratwick, "Alexandria, Syene, Meroe: Symmetry in Eratosthenes' Measurement of the World," in L. Ayres, ed., *The Passionate Intellect:*

Essays in the Transformation of Classical Traditions, New Brunswick: Transaction Publishers, 1995. See also: Aubrey Diller, "The Ancient Measurements of the Earth," *ISIS* 40, 1949, pp. 6–9; and J. B. Harley and D. Woodward, *The History of Cartography,* Vol. 1, Chicago: University of Chicago Press, 1987, pp. 148–160.

4.   Aelius Aristides, cited in Gratwick, p. 183.

5.   Cordell K. K. Yee, "Taking the World's Measure: Chinese Maps Between Observation and Text," in J. B. Harley and D. Woodward, eds., *The History of Cartography,* Vol. 2, Book 2. Chicago: University of Chicago Press, 1994, pp. 96–127 at p. 97.

6.   Pliny, *Natural History,* Book II, p. 247. The Loeb Classical Library, Cambridge, Mass.: Harvard University Press, 1997.

INTERLUDE: WHY SCIENCE IS BEAUTIFUL

1.   John Ruskin, ed., and abridged by D. Barrie. *Modern Painters.* Great Britain: Ebenezer Baylis & Son, 1967, p. 17.

2.   In their book *The Concepts of Science,* for instance, Lloyd Motz and Jefferson Weaver mention the occasional beauty of the field but warn that the "exaltation of our emotions and sensations to the level of great truths obscures the nature of scientific truth and opens the door to mysticism and metaphysics, which have no place in science." New York: Plenum, 1988, p. 12.

3.   Willa Cather, "Portraits and Landscapes," quoted in Daniel Halpern, ed., *Writers on Artists.* San Francisco: North Point Press, 1988, p. 354.

4.   Plato, *The Republic of Plato,* tr. Alan Bloom. New York: Basic Books, 1969, 605b.

5.   St. Augustine, *Confessions,* tr. R. S. Pine-Coffin. Baltimore, Md.: Penguin, 1970, Book X, section 33.

6.   Gottlob Frege, "On Sense and Reference," in *Translations from the Philosophical Writings of Gottlob Frege,* eds. Peter Geach and Max Black. Oxford: Blackwell, 1952, p. 63.

7.   On the distinction between beauty and elegance, see Michael Polanyi, "Beauty, Elegance, and Reality in Science," in S. Korner, *Observation and Interpretation in the Philosophy of Physics.* New York: Dover, 1957, pp. 102–106.

8.   In the words of philosopher Robin Collingwood: "The question has sometimes been raised, whether beauty is 'objective' or 'subjective,' by which is meant, whether it belongs to the object and is by it imposed on the mind by brute force, or whether it belongs to the mind and is by it

imposed on the object irrespective of the object's own nature. . . . [R]eal beauty is neither 'objective' nor 'subjective' in any sense that excludes the other. It is an experience in which the mind finds itself in the object, the mind rising to the level of the object and the object being, as it were, preadapted to evoke the fullest expression of the mind's powers. . . . Hence arises that absence of constraint, that profound sense of contentment and well-being, that characterizes the experience of real beauty. We feel that it is 'good for us to be here'; we are at home, we belong to our world and our world belongs to us." (R. G. Collingwood, *Essays in the Philosophy of Art.* Bloomington, Ind.: Indiana University Press, 1966, pp. 87–88.)

CHAPTER 2: DROPPING THE BALL:
THE LEGEND OF THE LEANING TOWER

1. See NASA's Lunar Feather Drop home page (http://vesuvius.jsc.nasa.gov/er/seh/feather.html).

2. Stillman Drake, *Galileo Studies: Personality, Tradition, and Revolution.* Ann Arbor, Mich.: University of Michigan Press, 1970, pp. 66–69.

3. Viviani, Vincenzio. *Vita di Galileo.* Milan: Rizzoli, 1954.

4. Cited in I. Bernard Cohen, *The Birth of a New Physics.* New York: Norton, 1985, p. 7.

5. *Ibid.,* pp. 7–8.

6. Quoted in Thomas B. Settle, "Galileo and Early Experimentation," in *Springs of Scientific Creativity: Essays on Founders of Modern Science,* eds. R. Aris, H. Davis, and R. Stuewer. Minneapolis: University of Minnesota Press, 1983, p. 8.

7. Galileo Galilei, *On Motion and on Mechanics,* tr. Stillman Drake, ed. I. E. Drabkin. Madison: University of Wisconsin Press, 1960.

8. Galileo Galilei, *Two New Sciences,* tr. Stillman Drake. Madison: University of Wisconsin Press, 1974, pp. 66, 75, 225–226.

9. Michael Segre, *In the Wake of Galileo.* New Brunswick, N.J.: Rutgers University Press, 1991, p. 111.

10. Christopher Hibbert, *George III: A Personal History.* New York: Basic Books, 2000, p. 194.

11. Gerald Feinberg, "Fall of Bodies Near the Earth," *American Journal of Physics* 33 (1965), pp. 501–503.

12. Thomas B. Settle, "Galileo and Early Experimentation," *op. cit.,* pp. 3–21.

13.   Stillman Drake, *Galileo at Work: His Scientific Biography*. Chicago: University of Chicago Press, 1978. See also Michael Segre, "Galileo, Viviani and the Tower of Pisa." *Studies in the History and Philosophy of Science* 20 (1989), 435–451. I am indebted to Thomas Settle for help with this chapter and the next.

INTERLUDE: EXPERIMENTS AND DEMONSTRATIONS

1.   Frederic Holmes, *Meselson, Stahl, and the Replication of DNA: A History of "The Most Beautiful Experiment in Biology."* New Haven: Yale University Press, 2001, pp. ix–x. "The [Meselson-Stahl] experiment originated in complexity, was surrounded by complexity, and directed the way toward the discovery of future complexities."

CHAPTER 3: THE ALPHA EXPERIMENT:
GALILEO AND THE INCLINED PLANE

1.   Galileo, *Two New Sciences*, tr. S. Drake, *op. cit.*, pp. 169–170.
2.   Alexandre Koyré, "An Experiment in Measurement," *Proc. American Philosophical Society* 97 (1953), pp. 222–236.
3.   Thomas B. Settle, "An Experiment in the History of Science," *Science* 133 (1961), pp. 19–23.
4.   Stillman Drake, *Galileo at Work: His Scientific Biography*. Chicago: University of Chicago Press, 1978, Ch. 5.

INTERLUDE: THE NEWTON-BEETHOVEN COMPARISON

1.   Owen Gingerich, ed., *The Nature of Scientific Discovery*. Washington, D.C.: Smithsonian Institution, 1975, p. 496.
2.   I. Bernard Cohen, *Franklin and Newton*. Philadelphia: American Philosophical Society, 1956, p. 43.
3.   Immanuel Kant, *Critique of Judgment*, tr. W. Pluher. Indianapolis: Hackett, 1987, section 47.
4.   Owen Gingerich, "Circumventing Newton: A Study in Scientific Creativity," *American Journal of Physics* 46 (1978), pp. 202–206.
5.   Jean-Marc Lévy-Leblond, "What If Einstein Had Not Been There? A *Gedankenexperiment* in Science History." 24th International Colloquium on Group-Theoretical Methods in Physics, Paris, July 2002.

CHAPTER 4: *EXPERIMENTUM CRUCIS:*
NEWTON'S DECOMPOSITION
OF SUNLIGHT WITH PRISMS

1.	I. Newton to H. Oldenburg, January 18, 1672, in W. Turnbull, ed., *The Correspondence of Isaac Newton,* Vol. I, Cambridge: University Press, 1959, pp. 82–83.

2.	Michael White, *Isaac Newton: The Last Sorcerer.* Reading, Mass.: Addison-Wesley, 1997, p. 165.

3.	Richard S. Westfall, "Newton," in *Encyclopaedia Britannica,* Fifteenth Edition, Vol. 24, p. 932.

4.	Quoted in White, *Isaac Newton: Last Sorcerer, op. cit.,* p. 179.

5.	*Ibid.,* p. 164.

6.	Newton, *Correspondence,* Vol. I, p. 92.

7.	Newton, *Correspondence,* Vol. II, p. 79.

8.	Newton, *Correspondence,* Vol. I, p. 107.

9.	*Ibid.,* p. 416.

10.	Thomas Birch, *The History of the Royal Society of London,* Vol. 3. New York: Johnson Reprint Corp., 1968, p. 313.

11.	Newton, *Correspondence,* Vol. I, p. 356.

INTERLUDE: DOES SCIENCE DESTROY BEAUTY?

1.	Kenneth Clark, *Landscape into Art.* New York: Harper & Row, 1976, p. 65.

2.	The fascinating variety of ways in which poets met this challenge has been discussed by, among others, Marjorie Nicolson in *Newton Demands the Muse: Newton's Opticks and the Eighteenth Century Poets,* Hamden, Ct.: Archon, 1963; and in M. H. Abrams, *The Mirror and the Lamp: Romantic Theory and the Critical Tradition.* New York: Oxford University Press, 1971.

3.	An entire book has been written about this party: Penelope Hughes-Hallett, *The Immortal Dinner: A Famous Evening of Genius and Laughter in Literary London.* Chicago: New Amsterdam, 2002.

4.	Nicolson, p. 25.

5.	From "The Best Mind Since Einstein," *NOVA,* November 21, 1993.

CHAPTER 5: WEIGHING THE WORLD:
CAVENDISH'S AUSTERE EXPERIMENT

1.	George Wilson, *Life of the Hon. Henry Cavendish.* London, Cavendish Society, 1851, p. 166. For a modern biography, see Christa Jungnickel and Russell McCormmach, *Cavendish: The Experimental Life*, Lewisburg, Penn.: Bucknell University Press, 1999.

2. *Ibid.*, p. 170.

3. *Ibid.*, p. 188.

4. *Ibid.*, p. 185.

5. *Ibid.*, p. 178.

6. Isaac Newton, *The Principia: Mathematical Principles of Natural Philosophy*, tr. I. Bernard Cohen and Anne Whitman. Berkeley: University of California Press, 1999, p. 815.

7. Isaac Newton, *Sir Isaac Newton's Mathematical Principles of Natural Philosophy and his System of the World*, tr. A. Motte, rev. by F. Cajori, Vol. 2. New York: Greenwood Press, 1969, p. 570.

8. In Derek Howse, *Nevil Maskelyne: The Seaman's Astronomer*. Cambridge: Cambridge University Press 1989, pp. 137–138.

9. Quoted in Russell McCormmach, "The Last Experiment of Henry Cavendish," A. Kox and D. Siegel, eds., *No Truth Except in the Details*. Dordrecht: Kluwer, 1995, pp. 13–14.

10. Henry Cavendish, "Experiments to Determine the Density of the Earth," *Philosophical Transactions of the Royal Society*, 88 (1798), pp. 469–526.

11. Quoted in B. E. Clotfelter, "The Cavendish Experiment as Cavendish Knew It," *American Journal of Physics* 55 (1987), 210–213, at p. 211.

12. Wilson, *op. cit.*, p. 186.

INTERLUDE: INTEGRATING SCIENCE
AND POPULAR CULTURE

1. Sarah Boxer, "The Art of the Code, or, At Play with DNA," *The New York Times*, March 14, 2003, p. E35.

CHAPTER 6: LIGHT A WAVE:
THOMAS YOUNG'S LUCID ANALOGY

1. The phrase "ontological flash" is from Mary Gerhart and Allan M. Russell, *Metaphoric Process: The Creation of Scientific and Religious Understanding*. Fort Worth: Texas Christian University Press, 1984, p. 114.

2. These and most other details of Thomas Young's life come from George Peacock, *Life of Thomas Young*, London: J. Murray, 1855, and the entry on Thomas Young by Edgar Morse in *The Dictionary of Scientific Biography*, Vol. 14. New York: Scribner's, 1976, pp. 562–572.

3.   Newton, *Opticks, op. cit.*, Query 28.

4.   Thomas Young, "Outlines of Experiments and Inquiries Respecting Sound and Light," *Philosophical Transactions 1800*, pp. 106–150.

5.   J. D. Mollon, "The Origins of the Concept of Interference," *Philosophical Transactions of the Royal Society of London* A (2002), pp. 360, 807–819. Newton's discussion is in *The Principia*, Book 3, proposition 24.

6.   Here, then, is one of the most obscure and backhanded introductions in the history of science of a fundamental concept: "It is surprising that so great a mathematician as Dr. Smith could have entertained for a moment the idea that the vibrations constituting different sounds should be able to cross one another in all directions, without affecting the same individual particles of air by their joint forces: undoubtedly they cross, without disturbing each other's progress; but this can be no otherwise affected than by each particle's partaking of both motions." Young, "Outlines," section 11.

7.   Thomas Young, *A Reply to the Animadversions of the Edinburgh Reviewers*. London: Longman et al. Cadell & Davis, 1804.

8.   Thomas Young, "The Bakerian Lecture: Experiments and Calculations Relative to Physical Optics." *Philosophical Transactions 1804*, pp. 1–16.

9.   Thomas Young, *A Course of Lectures on Natural Philosophy and the Mechanical Arts*. London: Taylor and Walton, 1845, lecture 39.

10.   Nahum Kipnis, *History of the Principle of Interference of Light*, Boston: Birkhäuser, 1991, p. 124.

11.   Henry Brougham, "Bakerian Lecture on Light and Colors," *The Edinburgh Review* 1 (1803), pp. 450–456.

INTERLUDE: SCIENCE AND METAPHOR

1.   This section is based on a column in *Physics World* ("Physics, Metaphorically Speaking," November 2000, p. 17) that, in turn, was partly drawn from Chapter Three of my *The Play of Nature: Experimentation as Performance*.

2.   Science historian Stanley Jackson has shown, for instance, that Johannes Kepler, like many scientists of the late sixteenth and early seventeenth centuries, projected a secular version of a soul-like animistic force into his mechanics. "If we substitute for the word 'soul' the word 'force' then we get just the principle which underlies my physics of the skies," Kepler wrote in 1621. Although he now rejected the idea that such a force was spiritual, he added, he had come "to the conclusion that this force must

be something substantial—'substantial' not in the literal sense but . . . in the same manner as we say that light is something substantial, meaning by this an unsubstantial entity emanating from a substantial body."

3. These are what philosopher Bruce Wilshire calls "physiognomic metaphors."

4. Finally, some scientific terminology looks metaphorical but isn't. Examples are the quark names "charm," "strange," "beauty," and "truth." It is just silly to think of these as metaphors. Such names do not tell us anything and aren't attempts to get at what something is. They are just ways of being irreverent.

5. One illustration of why it is important to understand this process is the recent "science wars," much of which revolved around whether the metaphors one encounters in science are creative (hence implying that the knowledge generated is culturally and historically bound) or filtrative (hence discardable). An example is the exchange about relativity between sociologist Bruno Latour, professor at the Centre de Sociologie de l'Innovation at the École Nationale Supérieure des Mines in Paris, and John Huth. Latour, examining a book in which Einstein relied on the imagery of observers and measuring sticks to explain relativity, argued that that was a sign the theory was socially constructed. Huth noted that the book was a popularization, pointed out that the imagery was inessential to the theory Einstein was trying to explain, and dismissed Latour's method as "metaphor mongering." (John Huth, "Latour's Relativity," *A House Built on Sand*, ed. N. Koertge, New York: Oxford University Press, 1998, pp. 181–192.) That metaphor might play a yet deeper role is suggested by Peter Galison, a historian of science at Harvard University, who pointed out the importance to Einstein's thinking of clock-synchronization methods in turn-of-the-century Europe to coordinate trains. Not only was Einstein familiar with the technology as a patent officer, Galison argues, but it helped lead him, by metaphorical extension, to solve the problem of simultaneity given the finite velocity of light. Einstein's key insight, according to Galison, was to drop the requirement that there be a "master clock."

CHAPTER 7: SEEING THE EARTH ROTATE:
FOUCAULT'S SUBLIME PENDULUM

1. "The Foucault Pendulum" (no author), *The Institute News*, April 1938.

2. Quoted in Stephane Deligeorges, *Foucault et ses Pendules*. Paris: Éditions Carre, 1990, p. 48.

3. M. L. Foucault, "Physical Demonstration of the Rotation of the Earth by Means of the Pendulum," *Journal of the Franklin Institute*, May 1851, pp. 350–353.

4. M. L. Foucault, "*Démonstration expérimentale du mouvement de rotation de la Terre*," *Journal des Débats*, 31 March 1851. For more on Foucault, see Amir Aczel, *Pendulum: Leon Foucault and the Triumph of Science.* New York: Pocket Books, 2003; William John Tobin, *The Life and Science of Léon Foucault, the Man Who Proved the Earth Rotates.* London: Cambridge University Press, 2003.

5. Demonstrating the translational movement of the earth—its movement through space rather than its rotation on its axis—would be harder.

6. M. Merleau-Ponty, *Phenomenology of Perception*, tr. C. Smith. London: Routledge & Kegan Paul, 1962, p. 280. I am grateful to Patrick Heelan for some of the following thoughts, and to Bob Street, who wondered what would happen if a person were fit into the suitably large bob of a pendulum, or if a pendulum were installed in a revolving restaurant whose period were a sidereal day, and seen against the background of a clear night sky.

7. Deligeorges, *Foucault*, p. 60.

8. H. R. Crane, "The Foucault Pendulum as a Murder Weapon and a Physicist's Delight," *The Physics Teacher*, May 1990, pp. 264–269, at p. 269.

9. H. R. Crane, "How the Housefly Uses Physics to Stabilize Flight," *The Physics Teacher*, November 1983, pp. 544–545.

10. The critical components are cable, the cable mount, and a little persuader in the collar of the cable mount. This device—a version of the one Foucault used but installed at the top of the pendulum rather than at the bottom—gives the cable a little push from time to time to keep the pendulum from slowing down.

11. What differs among the various pendulums is how much their direction of swing changes per hour, which is a function of their location. At the North and South poles, the pendulum would make a full circuit—360 degrees—every twenty-four hours, moving 15 degrees per hour, clockwise in the northern hemisphere, counterclockwise in the southern. Elsewhere, the hourly change depends on the latitude in a way specified as follows: The hourly deviation is 15 degrees times the sine of the latitude. In London, it is just shy of 12 degrees; in Paris, 11 degrees per hour; New York, 9¾ degrees per hour; New Orleans, 7 degrees, Sri Lanka, less than 2 degrees an hour.

INTERLUDE: SCIENCE AND THE SUBLIME

1.   Edmund Burke, "A Philosophical Inquiry into the Origin of Our Ideas of the Sublime and the Beautiful," 4th ed. Dublin: Cotter, 1707, Part 1, Section 6.
2.   Immanuel Kant, *Critique of Judgment*, tr. W. Pluher. Indianapolis: Hackett, 1987, Section 28. Yet another kind of sublimity is conveyed in Umberto Eco's novel *Foucault's Pendulum*, tr. William Weaver. New York: Ballantine Books, 1988.

CHAPTER 8: SEEING THE ELECTRON:
MILLIKAN'S OIL-DROP EXPERIMENT

1.   The best all-around article on Millikan's experiment remains Gerald Holton's seminal "Subelectrons, Presuppositions, and the Millikan-Ehrenhaft Dispute," in *The Scientific Imagination: Case Studies* (Cambridge, Mass.: Cambridge University Press, 1978) pp. 25–83; some of the other literature is discussed in Ullica Segerstråle, "Good to the last drop? Millikan Stories as 'Canned' Pedagogy," *Science and Engineering Ethics* 1:3 (1995), pp. 197–214.
2.   Millikan, *Autobiography*, New York: Houghton Mifflin, 1950, p. 69.
3.   *Ibid.*, p. 73.
4.   *Ibid.*
5.   Holton wrote that "Millikan did not design or devise the experiment from which his early fame sprang; rather, he discovered the experiment. . . . No one had doubted the existence of individual droplets. Anyone could have put together existing equipment a good decade or more earlier if one had only thought of watching a drop instead of a cloud. . . . The stranglehold on the imagination exerted by the tradition of work on clouds appears to have yielded only to Millikan's accident." (Holton, *op. cit.*, p. 46.)
6.   *Ibid.*, p, 53.
7.   Millikan, *Autobiography*, p. 75.
8.   *Ibid.*, p. 83.
9.   Millikan, "The Isolation of an Ion, a Precision Measurement of Its Charge, and the Correction of Stokes' law." *Science* 32 (1910), p. 436.
10.   The notebook page in question is reproduced in Holton, "Subelectrons," p. 64.
11.   Millikan, "On the Elementary Electrical Charge and the Avogadro Constant," *Physical Review* 2 (1911), pp. 109–143.
12.   Personal communication, Herbert Goldstein.

13. Holton, *op. cit.*, p. 71.

14. Segerstråle, *op. cit.*

15. Exposés have always had a powerful appeal, but especially in the post-Watergate period, when Holton's article appeared. Media critics like David Foster Wallace have examined why we "really like the idea of secret and scandalous immoralities unearthed and dragged into the light and exposed." Exposés, Wallace writes, give us the impression of "epistemological privilege," of "penetrating the civilized surface of everyday life" to reveal bad and seamy, even malevolent forces at work. David Foster Wallace, "David Lynch Keeps His Head," in *A Supposedly Fun Thing I'll Never Do Again* (New York: Little, Brown), p. 208.

16. Many other eminent scientists were also tarred by Broad and Wade, including Galileo. Taking Galileo's fictionalized dialogue to represent historical narrative, and relying on science historian Koyré's idiosyncratic interpretation of Galileo, Broad and Wade included him on their list of "known or strongly suspected cases of fraud in science" for having "exaggerated the outcome of experimental results." They buried in a footnote references to more recent historians of science who have examined Galileo's notebooks at length, such as Settle and Drake, and present convincing evidence that Koyré badly misunderstood Galileo.

17. A. Franklin, "Forging, Cooking, Trimming, and Riding on the Bandwagon," *American Journal of Physics* 52 (1984), pp. 786–793.

18. *Ibid.*, p. 83.

INTERLUDE: PERCEPTION IN SCIENCE

1. Cited in Evelyn Fox Keller, *Reflections on Gender and Science*. New Haven: Yale University Press, 1985, p. 165.

2. Anonymous, *Science News* 139 (1990), p. 359.

3. See Robert P. Crease, *The Play of Nature: Experimentation as Performance* (Bloomington, Ind.: Indiana University Press, 1993); Patrick A. Heelan, *Space-Perception and the Philosophy of Science* (Berkeley: University of California Press, 1983); Don Ihde, *Technology and the Lifeworld* (Bloomington, Ind.: Indiana University Press, 1990).

4. One important complication, however, is that a scientific term (such as "electron") can have what has been called a "dual semantics," since it can refer both to an abstract term in a theory and to a physical presence in a laboratory (consider the difference, for instance, between a "C" in a musical score and a "C" heard in a concert hall). On the dual seman-

tics of science, see Patrick A. Heelan, "After Experiment: Realism and Research," *American Philosophical Quarterly* 26 (1989), pp. 297–308, and Crease, *Play of Nature,* pp. 88–89.

5.   Albert Einstein, in Clifton Fadiman, ed., *Living Philosophies.* New York: Doubleday, 1990, p. 6.

CHAPTER 9: DAWNING BEAUTY: RUTHERFORD'S DISCOVERY OF THE ATOMIC NUCLEUS

1.   The classic article on this experiment is J. L. Heilbron, "The Scattering of α and β Particles and Rutherford's Atom," *Archive for History of Exact Sciences* 4 (1967), p. 247–307.

2.   M. Oliphant, *Rutherford: Recollections of the Cambridge Days.* Amsterdam: Elsevier, 1972, p. 26.

3.   J. A. Crowther, *British Scientists of the Twentieth Century.* London: Routledge & Kegan Paul, 1952, p. 44.

4.   A. S. Russell, "Lord Rutherford: Manchester, 1907–1919: A Partial Portrait," *Proceedings of the Physical Society* 64 (1 March 1951), p. 220.

5.   Oliphant, p. 123.

6.   Quoted in *Ibid.,* p. 65.

7.   J. L. Heilbron, "An Era at the Cavendish," *Science* 145, 24 August 1964, p. 825.

8.   Oliphant, *op. cit.,* p. 11.

9.   Quoted in D. Wilson, *Rutherford: Simple Genius.* Cambridge, Mass.: MIT Press, 1983, p. 290.

10.   E. N. da C. Andrade, *Rutherford and the Nature of the Atom.* New York: Doubleday, 1964, p. 111.

11.   Quoted in Wilson, *op. cit.,* p. 296.

12.   Quoted in A. S. Eve, *Rutherford.* New York: Macmillan, 1939, p. 199.

13.   Quoted in *Ibid.,* pp. 194–195.

14.   J. G. Crowther, "On the Scattering of Homogeneous Rays and the Number of Electrons in the Atom," *Proceedings of the Royal Society of London* 84 (1910–1911), p. 247.

15.   E. Rutherford, "The Scattering of α and β Rays and the Structure of the Atom," *Proceedings of the Manchester Literary and Philosophical Society,* series 4, 55, no. 1 (March 1911), p. 18.

16.   E. Rutherford, "The Scattering of α and β Particles by Matter and the Structure of the Atom," *Philosophical Magazine* (May 1911), pp. 669–688.

INTERLUDE: ARTISTRY IN SCIENCE

1. Quoted in Robert P. Crease and Charles C. Mann, *The Second Creation: Makers of the Revolution in Twentieth-Century Physics*. New Brunswick, N.J.: Rutgers University Press, pp. 337–338.
2. All the quotations in this paragraph are drawn from Patrick McCray, "Who Owns the Sky? Astronomers' Postwar Debates over National Telescopes for Optical Astronomy" (unpublished paper).
3. Robert P. Crease, *The Play of Nature: Experimentation as Performance*. Bloomington, Ind.: Indiana University Press, 1993, p. 109–111.
4. The exchange is quoted in Crease, *The Play of Nature*, pp. 117–118.

CHAPTER 10: THE ONLY MYSTERY: THE QUANTUM INTERFERENCE OF SINGLE ELECTRONS

1. R. P. Feynman, R. B. Leighton, and M. Sands, *The Feynman Lectures on Physics*, Vol. 3 (Menlo Park: Addison-Wesley, 1965), Chapter One; some of the following quotations are also drawn from Feynman, *The Character of Physical Law* (Cambridge, Mass.: MIT Press, 2001), Chapter Six.
2. Feynman's analogy, like all analogies, is only approximate, and on closer inspection is not as clean as it looks. Bullets may collide with one another before reaching the detector, which would alter the pattern. And if bullets as small as electrons ricochet from the edge of a tiny screen, they (unlike real bullets) suffer and impart to the screen a change in momentum, which might affect the pattern and the next interaction between an incoming electron and the screen. Finally, Feynman's contrast of bullets with water waves is for rhetorical effect. When we dilute any kind of matter, it eventually becomes atoms or fields, which are both quantized—so we *never* get a continuous wavelike pattern.
3. From 1888 to 1973, the Physical Institute was situated in the center of town, and physicists working on high-resolution electron microscopy or electron interferometry in the latter years had to struggle with mechanical and magnetic disturbances arising from urban life. In 1973, the Institute moved into new buildings atop a hill outside of town. Just as astronomers want their telescopes built far away from the lights of civilization, so Möllenstedt wanted his Institute far away from electromagnetic disturbances.

4.   He proceeded as follows: As a temporary substrate he used a 4×4 cm. glass plate coated with a thin 20-nanometer silver layer deposited by evaporation. This was thick enough for an electroplating with copper to a 0.5-micrometer-thick layer of foil. But how to get the small slits into the foil? His first idea was to scratch them into it with a scratching machine, as was the practice in producing light-optical interference grids. But no scratching machine was handy, and it appeared very difficult to scratch slits only 0.5 micrometers long with such a machine (such a short length was necessary to make the foil mechanically stable). This was where Jönsson's old electroplating experiments came in. Remembering that the smallest amount of dirt on the substrate prevents the growth of an electroplating layer, he put slit-shaped insulating layers on the silver substrate before electroplating. Now another of Möllenstedt's maxims became important: If you discover you have an effect due to dirt in your experiment, try to make it work for you. Jönsson found he did indeed have an effect due to dirt in the form of so-called Steward layers, which arise from oil molecules condensed from oil vapor inside the electron microscope. These oil molecules were "cracked" by the electron beam and polymerized to form the Steward layer. The longer a person took to look at an object, the thicker grew the Steward layer, reducing the contrast of the image. Jönsson experimented with Steward layers and found that they were good insulators, preventing the electroplating of copper at the spots on the silver substrate where they condensed. I am grateful to Claus Jönsson for help with these explanations of his experiments.

5.   He constructed an electron-optical device to produce an electron-probe to print a slit-shaped Steward layer on the silver substrate. To print several slits (up to ten) side by side, he supplied his apparatus with a condenser to divert the electron-probe by electric voltages vertical to the slit direction. After establishing the exposure time to get the 10–50-nanometer thick Steward layers, Jönsson was able to produce the desired slits in the copper layer. But how to remove them from the substrate, and how to remove the silver and polymerizate out of the slits? Here nature gave Jönsson a boost. He noticed that he could use a pair of tweezers to take the copper-silver-foil in slit direction off the glass plate without destroying the slits. When he prepared the slits over the 0.5-micrometer hole of a hole diaphragm and looked at them by a microscope, he saw that there was no matter in the slits. During the printing process, the electron beam bound the Steward layers on the

glass-silver-substrate, where they remained even when the copper foil was removed. Two big problems in the preparation of the slits thus proved not to be a hurdle.

6. In 1972, they obtained the first interference fringe patterns with an electron biprism inserted in the specimen cartridge of a Siemens Elmiskop 1A electron microscope equipped with a custom-made pointed filament. This work was awarded a prize for the best teaching experiment from the Italian Physical Society.

7. P. G. Merli, G. F. Missiroli, and G. Pozzi, *American Journal of Physics* 44 (1976), pp. 306–307.

8. The Web address is www.bo.imm.cnr.it.

9. A. Tonomura, J. Endo, T. Matsuda, T. Kawasaki, and H. Ezawa, "Demonstration of Single-Electron Buildup of an Interference Pattern," *American Journal of Physics* 57 (1989), pp. 117–120.

10. Tonomura's talk at the Royal Institution is available at http://www. vega.org.uk/series/vri/vri4/index.php. See also Peter Rodgers, "Who Performed the Most Beautiful Experiment in Physics?" *Physics World,* Sept. 2002.

INTERLUDE: RUNNERS-UP

1. Our ancient source for this story is the Roman architect and engineer Vitruvius Pollio, and the story here is quoted from the entry "Archimedes" in the *Dictionary of Scientific Biography,* Charles C. Gillispie, editor in chief. New York: Scribner, 1970–1980.

2. Frederic Lawrence Holmes, *Meselson, Stahl, and the Replication of DNA: A History of "The Most Beautiful Experiment in Biology."* New Haven: Yale University Press, 2001.

3. See Deborah Blum, *Love at Goon Park: Harry Harlow and the Science of Affection.* Cambridge, Mass.: Perseus, 2002.

4. J. Garcia and R. Koelling, "Relation of cue to consequence in avoidance learning," *Psychonomic Science* 4 (1966), pp. 123–124.

5. Available on the Web at http://www.aps.org/apsnews/0101/010106. html.

6. This is borrowed from the nontechnical description in Robert P. Crease and Charles C. Mann, *The Second Creation: Makers of the Revolution in Twentieth-Century Physics.* New York: Macmillan, 1986, pp. 164–165.

7. For a nontechnical description, see *Ibid.,* pp. 206–208.

8. For a nontechnical description, see Robert P. Crease, *Making Physics: A Biography of Brookhaven National Laboratory, 1946–1972.* Chicago: University of Chicago Press, 1999, pp. 248–250.

9. Quoted in *Ibid.,* p. 400.

CONCLUSION:
CAN SCIENCE STILL BE BEAUTIFUL?

1. The precise value of the muon's anomalous magnetic moment, as one might guess from the number of times it has been measured despite the difficulty involved, is one of the most avidly sought-after numbers in physics. The reason is that any discrepancy between the theoretically calculated number and what experimenters actually measure would reveal vital information about what might lie beyond the standard model of elementary particle physics—the theoretical package, assembled in the second half of the twentieth century, that describes the behavior of the basic building blocks of matter, including all known particles and most of the forces that affect them. See William Morse, et al., "Precision Measurement of the Anomalous Magnetic Moment of the Muon." Proc. of the XVIII Inter. Conf. on Atomic Physics, H. Sadeghpour, E. Heller and D. Pritchard, eds., World Scientific Publishing, 2002.

2. All muons spin continually on an axis, and at the same rate. When one follows a circular path in a uniform magnetic field, this axis will precess or wobble. The frequency of this wobble is determined by its gyromagnetic ratio, or "g-factor." In classical physics, where a particle's mass occupies a well-defined location in space and time, the g-factor would be exactly one. When P.A.M. Dirac combined relativity with quantum mechanics, he calculated the g-factor as exactly two. But according to the famous Heisenberg uncertainty principle of quantum mechanics, not all the mass of a muon (or of any other subatomic particle) can be pinpointed, and it is shrouded by a halo of ghostly and short-lived virtual particles, which the muon is continually emitting and absorbing. This makes its g-factor slightly different from two. Attempts to calculate the first order correction gave infinity, until Feynman, Schwinger, and Tomonaga were able to calculate it in quantum electrodynamics to 2.002. The g-2 experiment measures the difference between the g-factor and 2, or g-2, to better than one part in a million. The size of this number is of vital importance to physicists, for it might reveal new, as yet undiscovered particles. This in turn may

reveal whether or not the standard model is comprehensive. If the experimentally measured number hits the theoretically calculated number dead on, it would mean that the standard model is indeed comprehensive (at least for present purposes), and that any major new theory is probably a long way off. A discrepancy, however, would suggest that the standard model is not comprehensive, and would be a peephole into exciting new physics. To measure the wobble required planning and assembling an ingenious piece of equipment. To build it involved a decade-long effort that required supervising thousands of delicate pieces of equipment and getting them all to work together. It involved a huge number of trade-offs, for each part potentially affected all the others. The muons are created by a particle accelerator at Brookhaven called the AGS: Protons from the accelerator are slammed into a target, creating batches of other particles called pions, and these in turn decay into muons. These muons are polarized—their spin axes are all lined up in the same direction. Once inside a huge superconducting magnet, they are "kicked" so that they orbit in the center of the vacuum chamber inside the magnet. The magnet built at Brookhaven for this purpose is the largest single solid superconducting magnet in the world—so much bigger than its predecessors that some people thought the attempt would be hopeless. The field of this magnet has to be uniform and unchanging, and the scientists are constantly testing it for fluctuations. One method uses a special sensor-bearing trolley, built by the scientists, that makes periodic runs around the entire vacuum chamber; at one point, the scientists mounted a tiny video camera on it and made a video of the hour-long journey—like a trip through a long and very monotonous subway tunnel.

3. A muon decays into an electron (and two neutrinos), but the decay is not random; because of parity violation the high-energy electrons are spewed out in a preferred direction with respect to the muon's spin axis. These electrons are then picked up by detectors located along the inside of the ring.

4. A Cornell theorist named Toichiro Kinoshita had spent more than a decade grinding through the equations using the fastest computers available to produce high-order corrections to that number.

5. One way relativity is involved is that because the muons are traveling at nearly the speed of light they undergo time dilation, and instead of living 2.2 microseconds, they live a comparatively whopping 64 microseconds—a phenomenon that makes this experiment possible.

6. For a critique of the social constructivist approach to science—and the view that research is essentially a political or legal negotiation in which the parties swap interests—see Martin Eger, "Achievements of the Hermeneutic-Phenomenological Approach to Natural Science: A Comparison with Constructivist Sociology," in Robert P. Crease, ed., *Hermeneutics and the Natural Sciences*, Dordrecht: Kluwer, 1997, pp. 85–109.

7. It is tempting to take such "how we work" scenarios, as philosopher Maxine Sheets-Johnstone sarcastically dubs them, at face value. But they are only formalizations, and like any demonstration of a complex process they have been put together with purpose and ideology. The unspoken agenda of each approach to eliminate the body from science. Not, of course, the physical body of bones and blood, but what philosophers call the "lived body"—a primordial and unsurpassable unity productive of there being persons and worlds at all. The logic-oriented scholars want to eliminate this lived body, and to reconstruct science without its affective dimension, because it appears to introduce an element of arbitrariness and irrationality into what they see as an impersonal and objective process. It would be hard—and surely artificial—to find a place for beauty in such an account. Meanwhile, those scholars who focus exclusively on the social dimensions of science want to eliminate the lived body for the opposite reason—because admitting the foundational role of the animate human body into knowledge threatens to define generative and originary structures of human experience that would not only resist being reduced to social factors but indeed to some extent drive them—and can sweep them aside and even resist them. Little room exists for beauty here as well. For beauty is an intrinsic good, while the idiom of power struggle reduces all goods to instrumental goods. This approach to science therefore is just as dehumanizing as the logic-oriented approach, which depicts science in overly rational terms. For more on the role of the body in human inquiry, see Sheets-Johnstone, *The Primacy of Movement* (Philadelphia: John Benjamins, 1999). Like artists, scientists work with their entire beings, meaning that their work has an irreducibly affective dimension. If we strip science of its elements of affect and beauty we badly misrepresent it, and the result is a picture of science that is an academic figment, an artifact. A full account would involve a role for something like beauty—the definitive disclosure of what is fundamental so that we are held absorbed for a moment in the presence of something that

at once belongs to the realm of sense and the realm of ideas. A full account would also involve a role for love, the passion that is the correlate of the beautiful object: what a beautiful thing inspires, and what one feels for a beautiful thing.

8.   Plato, *Symposium*, 211C.

Index

About the Type

This book is set in Fournier, a typeface named for Pierre Simon Fournier, the youngest son of a French printing family. He started out engraving woodblocks and large capitals, then moved on to fonts of type. In 1736 he began his own foundry and made several important contributions in the field of type design; he is said to have cut 147 alphabets of his own creation. Fournier is probably best remembered as the designer of St. Augustine Ordinaire, a face that served as the model for Monotype's Fournier, which was released in 1925.